3 3073 00265805 0

DRUG ANALYSIS BY GAS CHROMATOGRAPHY

DRUG ANALYSIS BY GAS CHROMATOGRAPHY

David B. Jack

Department of Therapeutics and Clinical Pharmacology
The Medical School, Edgbaston
University of Birmingham
Birmingham, United Kingdom

1984

ACADEMIC PRESS, INC.
(Harcourt Brace Jovanovich, Publishers)
Orlando San Diego New York London
Toronto Montreal Sydney Tokyo

ACADEMIC PRESS, INC.
Orlando, Florida 32887

United Kingdom Edition published by
ACADEMIC PRESS INC. (LONDON) LTD.
24/28 Oval Road, London NW1 7DX

Library of Congress Cataloging in Publication Data

Jack, David B.
Drug analysis by gas chromatography.

Includes index.
1. Drugs--Analysis. 2. Gas chromatography.
I. Title. [DNLM: 1. Drugs--analysis. 2. Chromatography, Gas. QV 25 J115d]
RS189.J3 1984 615'.1901 84-9383
ISBN 0-12-378250-3 (alk. paper)

PRINTED IN THE UNITED STATES OF AMERICA

84 85 86 87 9 8 7 6 5 4 3 2 1

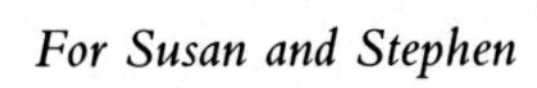
For Susan and Stephen

Contents

7. Measurement of Metabolites

8. Drug Screening

Preface

It may seem odd that some 20 years after gas chromatographs became generally available to laboratories engaged in drug analysis a book is appearing with the title "Drug Analysis by Gas Chromatography." Although the early days saw the publication of some excellent texts (e.g., "Gas Chromatographic Analysis of Drugs and Pesticides" by Gudzinowicz), there has been little since then to bridge the gap between elementary texts and research papers on specialized topics. In fact there have been very few reviews at all on the use of gas chromatography to measure drugs and excipients in pharmaceutical preparations.

One need only look at the current literature to see that gas chromatography is alive and well. It has, of course, been supplanted in some areas by high-performance liquid chromatography, particularly where very polar or high molecular weight drugs are concerned. However, the two techniques are often complementary, and, for many applications, gas chromatography is still the more sensitive. Electrochemical detection is only now reaching the limits that electron-capture detectors reached in the mid-1960s and is not nearly as stable. It is a fairly safe bet that gas chromatography will be around for a long time to come.

This book assumes that the reader is familiar with the fundamentals of gas chromatography. Chapter 1 introduces the column, the most important part of the gas chromatograph. There follows a chapter on derivatization techniques with examples using drugs, and Chapters 3 and 4 deal with the measurement of purity, stability, and the analysis of excipients and preservatives. Chapter 5 describes the determination of the active ingredients themselves. The next two chapters deal with the analysis of drugs and their metabolites in body fluids, and Chapter 8 describes the use of gas chromatography to identify the "unknowns" found in drug screening. Following the text, by way of an Envoi, is a reproduction of the article "Enlightenment" by G. Machata. Anyone who has worked in or managed a chromatography laboratory will empathize with this account. It deserves to be more widely available.

All too often, perhaps understandably, books and reviews concentrate on a narrow selection of journals for their material, and a conscious attempt has been made here to cast as wide a net as possible when the subject merits it. For this reason work from Dutch, Finnish, French, German, Japanese, Swedish, and Swiss sources has also been included. It will also serve as a reminder that good gas chromatography can, and is, carried out in non–English speaking countries.

I have not hesitated to draw upon my own experience, not from egotistical grounds but because this allows one to give the background as to *why* a particular approach was chosen. Unfortunately, today most papers do not give this type of information, and its omission is a sad loss.

Gas chromatography is still developing, and it is difficult to write an up-to-date account. The literature has here been covered up to the end of January, 1984, but I have not hesitated to include material that is much older if I considered it sound and appropriate. Some 2000 years ago Callimachus offered the opinion that a big book was a bad book, and I believe that, on balance, he was correct. I have tried to cover all the material relevant in a clear and concise manner.

There is a great deal of interesting work being carried out in gas–solid chromatography (little on drugs, however) that I have not included, and there are many interesting theoretical papers that I have considered outside the scope of this book. Anyone interested in the theory of gas chromatography who wishes to read some stimulating and original thinking on this topic could scarcely do better than to collect the papers published by L. S. Ettre or W. A. Aue and co-workers over the past 20 years.

I would like to thank Academic Press for their patience in allowing this work to come to fruition. It was begun among the wild and magnificent scenery of Eastern Azerbaijan with the encouragement of Professor Garnik Khatchaturian Santeh, University of Tabriz. I should like to thank him for his help during those happier times in a fascinating land.

David B. Jack

Chapter 1

Characterization of Stationary Phase and Drug

I. Introduction

Put at its simplest, gas chromatography allows the measurement of the time taken by a compound, a drug for example, to pass from the injection port to the detector. This measurement can be used to characterize both the stationary phase and drug, using the concept of retention index. Let us first begin by considering the retention volume V_R, which for any compound is a product of retention time and gas flow rate $V_R = t_R F_c$. This can be useful when comparing the retention of different compounds, especially when these retentions are measured on several different columns (i.e., on different stationary phases). Kovats (1958) introduced the concept of retention index to help in the comparison of different stationary phases. He chose the *n*-alkanes as standards, and defined the retention index of a compound I_x as follows:

$$I_x = 100\left[z + \frac{\log V_{Nx} - \log V_{Nz}}{\log V_{N(z+1)} - \log V_{Nz}}\right]$$

where z = the number of carbon atoms in the first *n*-alkane standard; V_{Nx} = the net retention volume of compound x; V_{Nz} = the net retention volume of the first *n*-alkane standard; and $V_{N(z+1)}$ = the net retention volume of the second *n*-alkane standard.

This rather complicated expression describes a very simple relationship; it relates the retention volume of a compound x to the retention volumes of the *n*-alkanes eluting from the column immediately before and after it. This

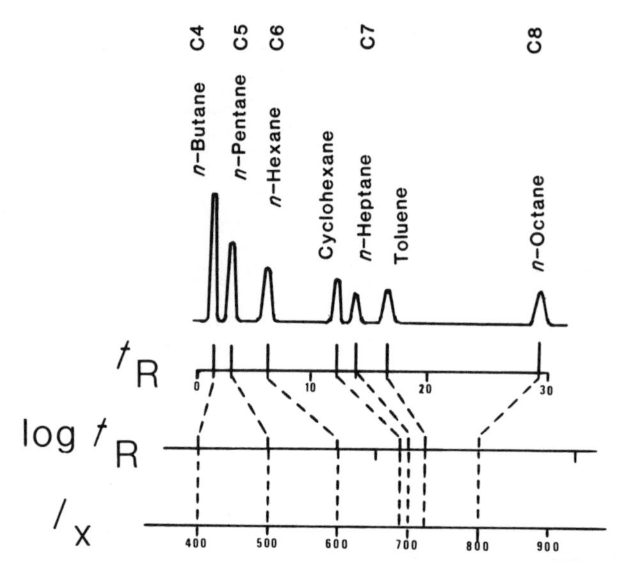

FIG. 1.1. Chromatogram of five standard alkanes and two other compounds, cyclohexane and toluene. From Ettre (1964). Reprinted with permission from the American Chemical Society.

can be illustrated simply by using the nomogram developed by Ettre (1964). It is given in a slightly modified form above.

Figure 1.1 shows a chromatogram of a mixture of five standard alkanes and two other compounds, cyclohexane and toluene. By definition in the Kovats system the *n*-alkanes that are shown have the following retention indices, directly related to their carbon numbers:

n-Alkane standard	Carbon no.	I_x
n-Butane	4	400
n-Pentane	5	500
n-Hexane	6	600
n-Heptane	7	700
n-Octane	8	800

From an examination of the chromatogram in Fig. 1.1, it can be seen that cyclohexane elutes from the column after *n*-hexane but before *n*-heptane. It will have, therefore, a retention index between 600 and 700. Similarly, since toluene elutes after *n*-heptane, but before *n*-octane, it will have a retention index between 700 and 800. The exact retention indices for cyclohexane and toluene can be obtained from the nomogram or by calculation from the formula given above.

Let us consider a practical example in the field of drug screening. Suppose that a patient is admitted to a hospital in a comatose state and with a very slow pulse. A drug screen is performed (see Chapter 8 for details), and the patient is found to have a blood alcohol level of 200 mg/100 ml, no acidic or neutral drugs, but a basic drug in his plasma. The high alcohol level could explain his comatose state but not his slow pulse. A mixture of straight-chain hydrocarbons is injected on to an SE-30 column followed by the basic plasma extract. The retention index of the basic compound is calculated to be 1865. Moffat's retention index data (Chapter 8, page 206) are scanned between 1845 and 1885. A number of drugs can be eliminated immediately because they would not be extracted into the basic fraction. The following possibilities are left:

Theobromine, 1847
Etryptamine, 1848
Phenazone, 1848
α-Meprodine, 1850
Neostigmine, 1850
Tymazoline, 1850
Trimeperidine, 1851
Benzphetamine, 1855
Dichlorophenazone, 1855
Lignocaine, 1870
Oxprenolol, 1870
Diphenhydramine, 1873
4-Hydroxyantipyrine, 1874
Orphenadrine, 1875
Dimethyltryptamine, 1885.

Of these, the drug oxprenolol is the most likely since it is a β-adrenoceptor antagonist (a β-blocker) and slows the heart. The rapid extraction of a urine specimen from the patient followed by thin-layer chromatography reveals the presence of a compound having the same R_f value as oxprenolol (Jack *et al.*, 1980). The retention index from Moffat's data is 1870. If further proof is needed, oxprenolol can be chromatographed on a column of different polarity or converted to a trifluoroacyl derivative (Chapter 6, page 124).

In order to eliminate possibilities with similar retention indices, other data such as UV absorption may also have to be used. Care should be exercised in using retention indices. Kazyak and Permisohn (1970) claimed that they were independent of temperature but Caddy *et al.* (1973) demonstrated a linear variation. However, the latter investigators studied the retention indices of their drugs at only two temperatures (150 and 190°C). Möller (1976) has published a study of the retention indices of 14 common drugs. Two different stationary phases were used (OV 1 and OV 17), and the work was carried out in three different laboratories on four gas chromatographs. A nonlinear variation in retention index was found over the four temperatures examined (150, 190, 230, and 270°C). The interlaboratory variation was reduced by one-third to one-half when comparisons were made at the same temperature.

II. The Column, the Support, and the Stationary Phase

A. The Column

The column is the heart of the gas chromatograph, and no matter how sophisticated the injection system and detector, this is what determines the success or failure of a separation. The column itself can be made of metal, glass, or teflon, the latter being restricted to operation at low temperatures and where no other column material is suitable. Metal columns are robust and can be used when it can be demonstrated that the compounds to be separated do not undergo decomposition on the hot metal surface. For the majority of drug analyses, however, this type of thermal decomposition is a problem and glass columns are advisable. Glass-lined metal columns are available that combine the inertness of the former with the sturdiness of the latter. For drug analysis, it is generally a good idea to silanize the empty column before packing it. Different types of glass columns can be used and these can be divided into the following classes: packed, micropacked, wall-coated open tubular, and support-coated open tubular.

1. Packed Columns

Normally, packed columns are between 1 and 4 m in length with an internal diameter of 2–4 mm. This means that their volume ranges from 3 to 50 cm^3. As the name suggests, the empty column is filled with packing: a chosen support coated with the desired amount of suitable stationary phase. The nature of the supports and phases available will be discussed shortly. Compared with the other types of column to be discussed, the efficiency of packed columns is low, 500–2000 theoretical plates. This is not always as important in drug analysis as it may be in some other fields, say petrochemicals, since complex mixtures are relatively rare. In measuring drugs in biological fluids, for example, the drugs will first be extracted by means of an organic solvent or other technique. This single step will substantially reduce the number of compounds present. Only in areas such as phytochemistry will very complex mixtures of closely related compounds need to be separated.

If carefully treated, packed columns will give excellent results, and this combined with their low cost accounts for their popularity. Depending on the detector used, volumes up to 10 μl can be injected and devices that allow greater volumes to be introduced on to the column are available. A wide range of column packing is available commercially and a 3% coating of stationary phase is the most popular for general applications. If desired, support and stationary phase can be bought separately, and the analyst can simply prepare any loading he wishes. Details of the techniques can be obtained in a number of books (Burchfield and Storrs, 1962; Purnell, 1962; Dal Nogare and Juvet, 1962).

2. *Micropacked Columns*

Micropacked columns are longer than packed columns but have a smaller internal diameter and more theoretical plates. Their length ranges from 5 to 25 m, with diameter 0.6–1 mm, and 50,000 plates have been obtained without the use of very high inlet pressures. Because of their small volumes, approximately 2–30 cm^3, micropacked columns are very economical in use of column packing, and this may be important when some of the more expensive stationary phases are being used.

3. *Wall-Coated Open Tubular Columns*

Wall-coated open tubular columns are very long (10–100 m) with small internal diameters (0.1–0.5 mm). These columns are also called capillary and (rarely now) Golay columns. In an open tubular column there is no support, the stationary phase is coated directly onto the inside wall of the column. It is possible to bond the stationary phase chemically to the wall of the column (this results in the column retaining its efficiency over a longer period), and the column can actually be washed with organic solvent if it becomes contaminated.

The thickness of the coating of stationary phase depends on the amount of material to be separated; a relatively thin coating would be 0.1–0.2 μm, a medium one 0.2–0.8, and a thick coating would be 0.8–1.2 μm. A thin coating provides less resistance to mass transfer in the stationary phase itself and a decreased analysis time. Thicker coatings, however, have some advantages such as allowing a larger sample capacity and reducing the possibility of an interaction between the injected sample and the capillary wall. Some work has been published on very thick liquid coatings (around 5 μm), and this is well worth reading (Ettre, 1983; Ettre *et al.*, 1983).

Formerly, the use of capillary columns meant that a very restricted volume of solvent could be injected. This was in the submicroliter range and was introduced by means of a stream splitter device. While this is, of course, still possible, there are now a number of injection systems available that allow microliter volumes to be introduced without splitting. Open tubular columns have very high efficiencies (10,000–100,000 theoretical plates) and are ideal for the separation of very complex mixtures, particularly those containing closely related compounds, e.g., homologous series. They can also be used for much simpler separations where high efficiency is desired, such as the measurement of nano- and subnanogram concentrations of drugs in biological extracts.

In the past, capillary columns were made of glass and were extremely fragile. Now fused silica columns are available and these are much more robust and are enjoying increasing popularity. It is possible to convert a conventional gas chromatograph, using packed columns, to a capillary instrument by modifying the injection system and introducing some "makeup" gas immediately before the detector. If this latter modification is not made, some efficiency will be lost as the peaks emerging from the column are "diluted" in the detector.

4. Support-Coated Open Tubular Columns

Support-coated open tubular columns combine the advantages of both packed and open tubular columns: high efficiency at moderate cost. They are made by filling a length of glass tubing, several mm in diameter, with column packing and then drawing out the tubing into a capillary. Sample volumes of 0.1 μl can be injected, and it is not necessary to use a stream splitter. These columns have been successfully used to separate drugs of abuse (Caddy *et al.*, 1973), but their use may decline with the advent of fused silica capillary columns.

B. The Support

The support is the material on which the stationary phase is coated and, ideally, it should combine a large surface area with chemical inertness and low absorption. It should also be capable of being packed uniformly into a column without breaking up easily. If the support were to break up readily, a wide range of particle sizes would result, with an increase in eddy diffusion and decrease in efficiency. Inevitably, a compromise must be made regarding low absorbtivity and inertness since a support that was completely inert and nonabsorptive would be impossible to coat with stationary phase. A large surface area is desirable since separation of a mixture takes place only in the stationary phase and not in the moving gas phase. Hence, by providing a large surface area, rapid equilibrium and efficient separation are possible. Many different supports are available, and often the same support is sold by different companies under different names. There now follows a short review of the types of supports readily available.

1. *Diatomaceous Earth Supports*

Diatomaceous earth supports were among the earliest used and are still the most common. Diatoms are microscopic algae and inhabit fresh and salt waters in all parts of the world. When these plants die, their skeletons, which are almost pure silica, sink to the bottom and form sedimentary beds. In several parts of the world geological activity has brought beds which were laid down millions of years before to the surface. Large deposits have been found in the western states of the United States, Sweden, and Germany. The German name for these deposits is kieselguhr.

Diatomaceous earth has a very large surface area : weight ratio but the tiny skeletons are too small and fragile to be used directly. They are first heated to a high temperature (1000–1500°C) either alone or in the presence of a flux of sodium carbonate. If no carbonate flux is added, the resulting material is pink because of the oxidation of trace minerals, and this material is sold under names such as Chromosorb P, Firebrick, Anakrom P, and Gas Chrom P. If carbonate flux is added, the silica is converted to crisobalite, and trace metals are transformed to colorless silicates. The white-colored material resulting from such a flux calcination is sold under a series of names, e.g., Chromosorb W, Celite, Anakrom U, and Gas Chrom S. The calcined diatomaceous earth is sieved into a number of fractions of different particle size called meshes. The most common grades sold are 80–100 mesh (0.18–0.15 mm particle diameter) and 100–120 mesh (0.15–0.12 mm), although even finer size grades are offered, e.g., 70–80, and 80–90. The two most commonly used series are the Chromosorb and Gas Chrom supports, and the properties of the different members of both groups will be briefly outlined.

2. *Chromosorb Supports*

Chromosorb W is white, with a relative nonabsorptive surface, which makes it very suitable as a support for chromatography of polar compounds. It is, however, easily broken down into smaller particles, and a more robust form, Chromosorb G, with a lower surface area and higher density, has been developed. Chromosorb P, obtained by calcination without a flux of sodium carbonate, is more absorptive than W or G and is used for the chromatography of nonpolar or moderately polar substances. Another support, Chromosorb A, has been produced for preparative gas chromatography, where milligram quantities are collected as they emerge from the column in very pure form. Chromosorb A resembles P in appearance but is closer to W in surface area and performance. It is a good support for high loadings of stationary phase such as are needed in preparative work. Where very low loadings are desirable, as in chromatographing high molec-

Table 1.1

Properties of Some Chromosorb Supports

Chromosorb	P	W	G	Glass beads
Appearance	Pink	White	Pearl	Transparent
Activity	Partially inert	Inert	Inert	Inert
Packed density (g/ml)	0.47	0.24	0.58[a]	1.4
Surface area (m^2/g)	4	1	0.5	0.2
Normal coating of stationary phase (%)	5–30	1–25	1–20	0.05–2
Theoretical plates/meter	800–2000	700–1500	1000–2000	200–500
Breakdown (%) after 5 min shaking	12	19	1	0

[a] Because of its higher density, a loading of 3% (w/w) on Chromosorb G is equivalent to a loading of 7% (w/w) on Chromosorb W.

ular weight compounds such as steroids, Chromosorb G is the most useful. The physical properties of the frequently used Chromosorbs are given in Table 1.1, with another support, glass beads, for comparison.

3. *Gas-Chrom Supports*

The product of flux calcination with Gas-Chrom supports is called Gas-Chrom S, and it has a surface area : weight ratio of about 1 m^2/g. The product of calcination without flux, Gas-Chrom R, has a much larger surface area : weight ratio (4 m^2/g). It is claimed that Gas-Chrom R has the highest efficiency of all diatomaceous supports. Because of its high adsorptivity, however, its use is restricted to the separation of nonpolar substances such as hydrocarbons. Both the Chromosorb and Gas-Chrom series of supports are available in a large number of mesh sizes.

4. *Modification of Diatomaceous Earth Supports*

Very often diatomaceous earth supports are treated chemically to make them more inert before they are coated with stationary phase. The active sites that could bind polar compounds are associated with hydroxyl groups attached to trace elements, such as iron or aluminum, and to the silicon atoms themselves. Washing the support with acid is successful in removing trace elements; the support is then washed until it is neutral and dried before coating with stationary phase. Supports treated in this way have the letters AW (acid washed) added to their name, e.g., Chromosorb P AW. Washing with alkali is also sometimes carried out in order to remove active catalytic sites and to reduce "tailing" of peaks, especially with basic drugs. In order to deactivate hydroxyl groups attached to the silicon atoms themselves, a special process called silanization is employed.

The two most commonly used silanizing agents are dimethyldichlorosilane (DMCS) and hexamethyldisilazane (HMDS), and they react with Si-hydroxyl groups in the following manner:

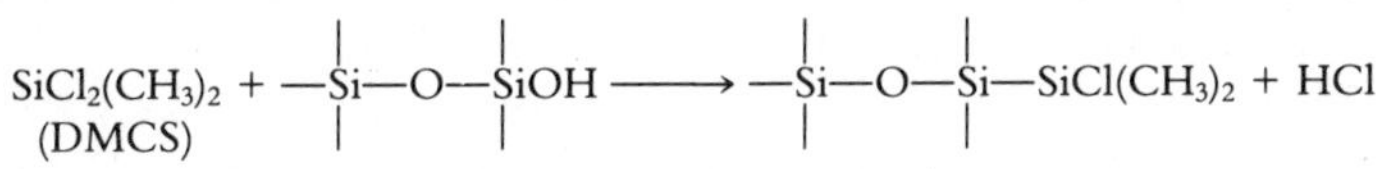

$$\underset{\text{(DMCS)}}{SiCl_2(CH_3)_2} + -\overset{|}{\underset{|}{Si}}-O-\overset{|}{\underset{|}{Si}}OH \longrightarrow -\overset{|}{\underset{|}{Si}}-O-\overset{|}{\underset{|}{Si}}-SiCl(CH_3)_2 + HCl$$

Or, if adjacent OH groups are present:

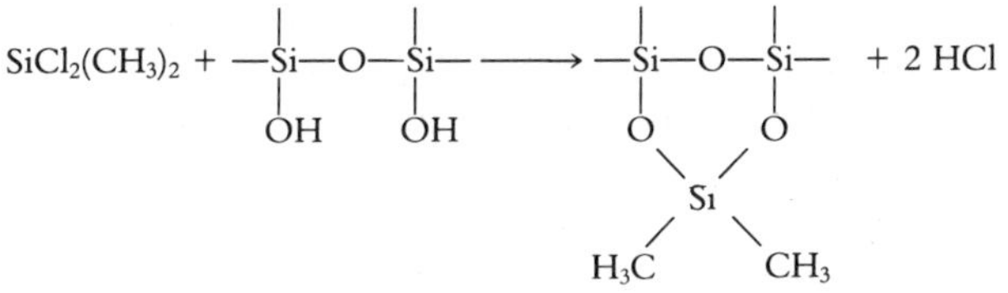

$$SiCl_2(CH_3)_2 + -\overset{|}{\underset{\underset{OH}{|}}{Si}}-O-\overset{|}{\underset{\underset{OH}{|}}{Si}}- \longrightarrow \begin{array}{c} -\overset{|}{\underset{\underset{O}{|}}{Si}}-O-\overset{|}{\underset{\underset{O}{|}}{Si}}- \\ \diagdown \qquad \diagup \\ Si \\ \diagup \qquad \diagdown \\ H_3C \qquad\quad CH_3 \end{array} + 2\,HCl$$

and

$$\underset{\text{(HMDS)}}{Si_2(CH_3)_6NH} + -\overset{|}{\underset{\underset{OH}{|}}{Si}}-O-\overset{|}{\underset{\underset{OH}{|}}{Si}}- \longrightarrow -\overset{|}{\underset{\underset{\underset{(CH_3)_3Si}{|}}{\underset{O}{|}}}{Si}}-O-\overset{|}{\underset{\underset{\underset{Si(CH_3)_3}{|}}{\underset{O}{|}}}{Si}}- + NH_3$$

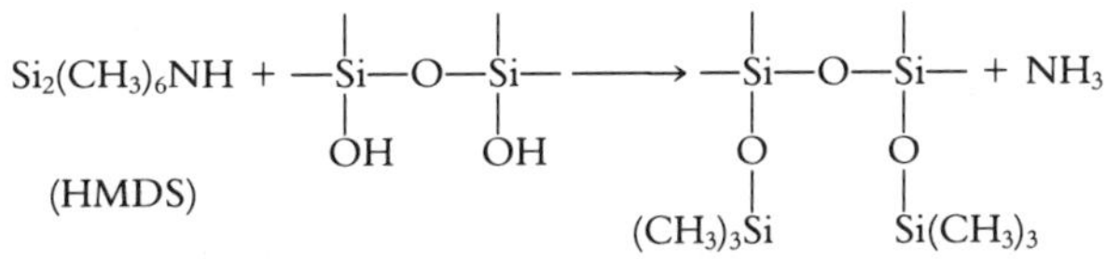

Where one DMCS molecule reacts with a single OH group, the remaining active Si—Cl linkage must itself be deactivated, and this is done by washing the silanized support with methanol when the following reaction takes place:

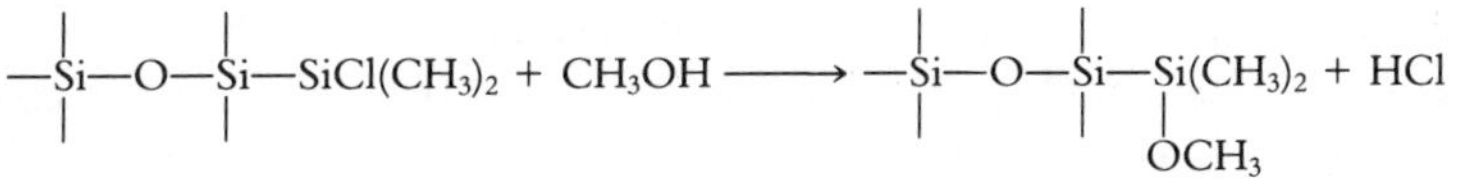

$$-\overset{|}{\underset{|}{Si}}-O-\overset{|}{\underset{|}{Si}}-SiCl(CH_3)_2 + CH_3OH \longrightarrow -\overset{|}{\underset{|}{Si}}-O-\overset{|}{\underset{|}{Si}}-\underset{\underset{OCH_3}{|}}{Si}(CH_3)_2 + HCl$$

Of course, the number and nature of the Si hydroxyl groups in any sample of diatomaceous earth are not known, and, in practice, washing with methanol is always carried out after silanization with DMCS. Since there is no reactive Si—Cl bond in HMDS, washing with methanol is not necessary when this reagent is used to deactivate supports. When a support has been commercially deactivated, it will carry the appropriate letters (e.g., Chromosorb G AW-DMCS).

5. Polymer Supports

In addition to the diatomaceous earth supports, there are a number of polymer supports of which the Porapak series is most frequently used. These supports are based on a polymer made up of styrene and ethylvinylbenzene units. A range of polarities are produced by modifying the degree of cross-linking in the polymer. Porapak P is the most nonpolar, and the order of increasing polarity is P, P (silanized), Q, Q (silanized), R, S, N, and T. These supports can be used up to a temperature of about 250°C and are sold as porous beads in several mesh sizes. The beads are used directly to fill a column and no stationary phase is needed. The Porapak supports are useful when very polar compounds are to be chromatographed, especially those that can be dissolved in aqueous or alcoholic solution. Fluorocarbon supports are also available under the names Chromosorb T (Teflon 6), Fluoropak 80, and Kel F. In order to be able to separate mixtures, the fluorocarbon supports must be coated with a suitable stationary phase, unlike the Porapak series. The efficiency of the fluorocarbon supports is low, but they are useful when very reactive substances are to be separated, e.g., halogen gases or hydrazines. Teflon-coated diatomaceous earth supports are also available under the names Gas Pack F and Gas Pack FS.

6. Other Supports

Glass beads are unreactive and allow short retention times because of the very regular packing of the small spheres. The percentage loading of stationary phase must be low because of the high density and low surface area of the beads. In general, low efficiencies are obtained, but this can be improved by etching the beads chemically before applying the stationary phase. A number of other materials have been used as supports, but none has been widely adopted for drug analysis. Graphitized carbon black (carbon black heated to several thousand degrees) has been used and so also has colloidal alumina (Boehmite). The latter shows great affinity for charged compounds such as carboxylic acids and chelating agents. An unusual support is the skeletal spicules of the marine sponge *Sterraster;* these are spheroidal in shape with a diameter of about 60 μm, and their surface is covered with an elevated texture, which provides an excellent surface for coating the stationary phase.

C. The Stationary Phase

To obtain a good separation of any mixture, it is essential that the stationary phase is correctly chosen. A good stationary phase should have a low vapor pressure, be thermally stable, and resistant to solvent attack. A low vapor pressure is essential to prevent the phase itself from being eluted from the column. Even with phases of low vapor pressure, operation at high column temperature results in some loss, which is called "bleeding," and is shown by a slow upward drift of the baseline displayed on the chart recorder.

Suppliers of gas-chromatographic materials list ~200 stationary phases in their catalogs. However, it is possible to classify phases in such a manner that they can be simply compared. There now follows a short discussion on the nature of the most frequently used phases, and this leads to an explanation of the use of the retention index to characterize stationary phases.

As we have seen earlier in this chapter, the separation of the components of a mixture takes place only in the stationary phase, and therefore the compounds to be separated must dissolve to some extent in the phase. It is a well-known chemical principle that "like dissolves like:" polar substances are more soluble in polar solvents while nonpolar substances dissolve more readily in nonpolar solvents. In the same way a polar phase will be better for separating a mixture containing polar constituents than a nonpolar one. The chemical structures of the most popular phases are given in Table 1.2.

The most outstanding feature of the phases shown is that the majority are polymeric in nature; this should not be surprising in view of the low vapor pressure and thermal stability required of a good phase. Many of the phases are highly viscous fluids, gums, or resins. The most widely used polymers are the polysiloxanes, which are synthesized in very pure form and have a common skeleton:

$$—\overset{|}{\underset{|}{Si}}—O—\overset{|}{\underset{|}{Si}}—O—\overset{|}{\underset{|}{Si}}—O—\overset{|}{\underset{|}{Si}}—O—\overset{|}{\underset{|}{Si}}—O—\overset{|}{\underset{|}{Si}}—O—\overset{|}{\underset{|}{Si}}—O—\overset{|}{\underset{|}{Si}}—O—$$

Different degrees of polarity are provided by varying the nature of the substituents attached to the silicon atoms. When all of the substituent groups are methyl, the phase is nonpolar, and such phases are sold under the names SE-30, OV-1, OV-101, and others. By introducing a number of phenyl groups, the polarity can be increased in a controlled manner. This can be illustrated by the very commonly used OV series (OV stands for Ohio Valley, where the polymers were originally produced). The series can be arranged in increasing polarity with the percentage of phenyl substitution

Table 1.2

Structures of Some Commonly Used Stationary Phases

Phase	Structure	Polarity[a]	Upper temperature limit
Squalane	H_3C, H_3C, CH_3, CH_3, CH_3, CH_3, H_3C, CH_3 (skeletal structure)	NP	150
SE-30	$-[Si(CH_3)_2-O-Si(CH_3)_2-O-Si(CH_3)_2-O]_n-$	NP	300
QF-1	$(CH_3)_3Si-[O-Si(CH_3)(C_2H_4CF_3)]_x-[O-Si(CH_3)_2]_y-O-Si(CH_3)_3$	IP	250
XE-60	$(CH_3)_3Si-[O-Si(CH_3)(C_2H_4C{\equiv}N)]_n-O-Si(CH_3)_3$	IP	250
NPGS	$-[OCH_2-C(CH_3)_2-CH_2-O-C(=O)-CH_2CH_2-C(=O)]_n-O-$	P	230
Carbowax	$HO-[CH_2CH_2O]_n-H$	P	100–225 (depending on MW)
DEGS	$-[C_2H_4-OC_2H_4-O-C(=O)-C_2H_4-C(=O)-O]_n-$	P	200

[a] NP, nonpolar; P, polar; IP, intermediate polarity.

given in parentheses: OV-1, (0%), OV-3, (10%), OV-7 (20%), OV-11 (35%), OV-17 (50%), OV-22 (65%) OV-25 (75%).

The polarity can be increased further by the introduction of substituents such as trifluoropropyl (OV-210 or QF-1), cyanoethyl (XE-60), or a combination of cyanopropyl, phenyl, and methyl (OV-225, OV-275). Dexsil 300-GC was developed by the Olin Corporation and is a carborane–siloxane polymer of unusual structure shown in Fig. 1.2. The makers claim efficient separations over a wide temperature range (50–500°C) with little bleeding. The polarity of this phase is greater than that of SE-30 but less than that of OV-25, and, if it becomes contaminated with the decomposition products of unstable compounds or with high molecular weight compounds that do not chromatograph, these can be eluted by raising the column temperature to 400°C.

Improvements are constantly being made in stationary-phase technology, and details of advances can be obtained from the literature or from the regular newsletters that most large suppliers publish. A detailed review of developments in polysiloxane stationary phases has been published and is well worth reading (Haken, 1984). Stationary phases that are capable of separating enantiomers are available, and these phases are optically active. Two principal types are in use: dipeptides like *N*-trifluoroacetyl-L-norvalyl-L-norvaline cyclohexyl ester and the carbonylbisaminoesters. Aue *et al.* (1973) demonstrated the preparation of a novel stationary phase. They subjected a normal 6% Carbowax 20M packing to exhaustive extraction with a polar solvent to leave a support with a very thin film of stationary phase, estimated to be only 15 Å thick. Such a packing lies on the border between gas–liquid and gas–solid chromatography. They were able to obtain an efficient separation of polar compounds using this packing; the amount of

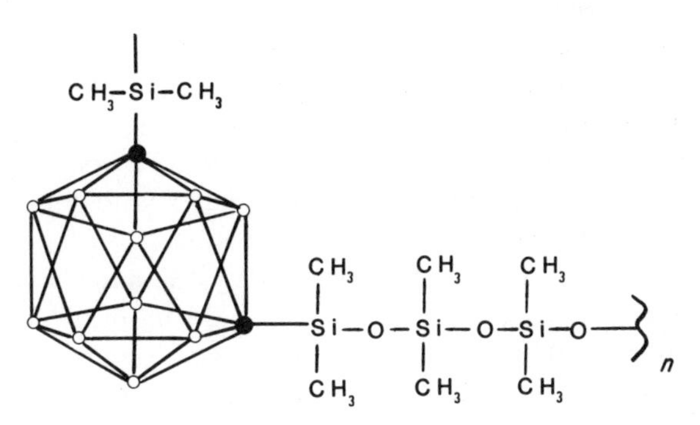

FIG. 1.2. Dexsil 300-GC. ●, Carbon; ○, B—H; $n = 50$.

Carbowax remaining on the support was estimated to be equivalent to a coating of 0.1–0.2%

D. Characterization of Stationary Phases Using Retention Index

At the beginning of this chapter the concept of retention index was introduced, and it was seen how compounds could be characterized by their retention indices. The same concept can also be adapted for the characterization of stationary phases. Rohrschneider (1966) selected a very nonpolar phase, squalane, as his standard stationary phase and compared the difference in retention index (ΔI) of a compound measured first on squalane and then on a more polar phase at the same temperature. Whence $\Delta I = I_{polar} - I_{squalane}$. Choosing five compounds of different polarity and different chemical properties (benzene, ethanol, methyl ethyl ketone, nitromethane, and pyridine), Rohrschneider considered the difference in retention index to be due to the products of two sets of constants: substance specific and phase specific:

$$\Delta I = ax + by + cz + du + es$$

where a, b, c, d, and e are the substance-specific and x, y, z, u, and s are the phase-specific constants. The values of both sets can be obtained by measuring retention indices on a large number of stationary phases and then solving a series of simultaneous equations of the type shown above. The structure-specific constants have been used to predict retention indices for a number of compounds, while the phase-specific constants have proved very useful in comparing stationary phases. Their use can be illustrated by reference to the constants obtained for a number of common phases (Table 1.3).

Three important points to note about using the Rohrschneider constants are (i) as the polarity of the stationary phase increases, so does the size of the constants, (ii) the greater the magnitude of a constant for a particular standard compound, the more that standard will be retarded on that phase (i.e., the longer will be its retention time), and (iii) phases with similar contants will behave in a chromatographically similar manner.

Thus since the first three phases, OV-1, OV-101, and SE-30, all have identical Rohrschneider constants, they will be indistinguishable chromatographically. Hence no useful purpose would be served by buying the three phases since any one would suffice. The increase in polarity in the OV series

Table 1.3

Rohrschneider Constants for Some Common Phases

	Constant				
Phase	*x*	*y*	*z*	*u*	*s*
OV-1, gum	0.16	0.20	0.50	0.85	0.48
OV-101, fluid	0.16	0.20	0.50	0.85	0.48
SE-30, gum	0.16	0.20	0.50	0.85	0.48
OV-3	0.42	0.81	0.85	1.52	0.89
OV-7	0.70	1.12	1.19	1.98	1.34
OV-11	1.13	1.57	1.69	2.66	1.95
OV-17	1.30	1.66	1.79	2.83	2.47
OV-25	1.76	2.00	2.15	3.34	2.81
OV-225	2.17	3.20	3.30	5.16	3.69

going from OV-1 to OV-225 is nicely illustrated by the parallel increase in constants.

A further advance was made by McReynolds in the United States in 1970. He substituted *n*-butanol, 2-pentanone, and nitropropane for ethanol, methyl ethyl ketone, and nitromethane because the retention times of the latter were very short on some stationary phases. In addition, he increased the number of standard compounds to 10 to improve the characterization of the phases and to extend the predictive use of retention indices. For example, the alcohol 2-methyl-2-pentanol was added to improve the prediction of retention indices for branched chain compounds, and 1-iodobutane was added to improve the prediction for halogen-containing compounds. The last three compounds he added, 2-octyne, 1,4-dioxane, and *cis*-hydrindane, gave only very minor improvements in the system.

Using these 10 compounds, McReynolds measured their retention indices on 226 different stationary phases, and, comparing them to their retention indices on squalane, calculated the ΔI values. Since their publication in 1970, the McReynolds constants have been widely adopted in the selection of suitable stationary phases for a particular separation. In arriving at his constants Rohrschneider had divided by a factor of 100; McReynolds did not do this, and consequently his constants are much larger. They are, however, used in exactly the same way. For example, if a laboratory wishes to use a published method where the phase used was XE-60 and finds none available, it may not be necessary to order this phase. An examination of the McReynolds constants shows that OV-225, while not exactly similar, is very close in character:

Standard compound	ΔI	
	XE-60	OV-225
Benzene	204	228
n-Butanol	381	369
2-Pentanone	340	338
Nitropropane	493	492
Pyridine	367	386
2-Methyl-2-pentanol	289	282
1-Iodobutane	203	226
2-Octyne	120	150
1,4-Dioxane	327	342
cis-Hydrindane	94	117

Thus OV-225 if available should be a reasonable substitute for XE-60 in most applications.

The constants can also be used in the initial choice of a phase. If a drug contains a particular group or atom, e.g., a double bond, is halogenated, or contains a carbonyl group, particular attention should be paid to the constants ΔI (benzene), ΔI (1-iodobutane), and ΔI (2-pentanone), respectively. These constants will indicate the degree of retardation that can be expected on a particular phase. Consider the example of the two phases OV-25 and QF-1.

Standard compound	ΔI	
	OV-25	QF-1
Benzene	138	144
n-Butanol	204	233
2-Pentanone	208	355
Nitropropane	305	463
Pyridine	280	305
2-Ethyl-2-pentanol	144	203
1-Iodobutane	169	136
2-Octyne	147	53
1,4-Dioxane	251	280
cis-Hydrindane	113	59

A drug containing an ester group (or other carbonyl-containing group) will be retarded more on QF-1 than on OV-25 because the ΔI for 2-pentanone on QF-1 is 355 and only 208 on OV-25. If, however, a drug contains a triple bond or is halogenated, it will be more retarded on OV-25 (compare the appropriate ΔI values for 2-octyne and 1-iodobutane). By employing the contants in this way their usefulness is readily apparent. The twenty-fifth anniversary of the use of the retention index system was celebrated in a review by Budahegyi *et al.* (1983).

Sometimes in drug analysis a less than ideal phase may be selected to give a more rapid elution. When drugs are to be measured in blood or similar fluid, a complex extraction is usually carried out, which results in a relatively clean extract where only drug and internal standard need be separated. Here a low polarity phase such as OV-1 often gives a rapid and satisfactory separation.

References

Aue, W. A., Hastings, C. R., and Kapila, S. (1973). *J. Chromatog.* **77,** 299.
Budahegyi, M. V., Lombosi, E. R., Lombosi, T. S., Meszaros, S. Y., Nyiredy, S., Tarjan, C., Timar, I., and Takacs, J. M. (1983). *J. Chromatog.* **271,** 213.
Burchfield, H. P., and Storrs, E. E. (1962). "Biochemical Applications of Gas Chromatography." Academic Press, New York and London.
Caddy, B., Fish, F., and Scott, D. (1973). *Chromatographia* **6,** 293.
Dal Nogare, S., and Juvet, R. S. (1962). "Gas-liquid Chromatography, Theory and Practice." Wiley, New York and London.
Ettre, L. S. (1964). *Anal. Chem.* **36**(8), 31A.
Ettre, L. S. (1983). *Chromatographia* **17,** 553.
Ettre, L. S., McClure, G. L., and Walters, J. D. (1983). *Chromatographia* **17,** 560.
Haken, J. K. (1984). *J. Chromatog.* **300,** 1.
Jack, D. B., Dean, S., and Kendall, M. J. (1980). *J. Chromatog.* **187,** 277.
Kazyak, L., and Permisohn, R. (1970). *J. Forensic Sci.* **15,** 346.
Kovats, E. (1958). *Helv. Chim. Acta* **41,** 1915.
McReynolds, W. O. (1970). *J. Chromatogr. Sci.* **8,** 685.
Möller, M. R. (1976). *Chromatographia* **9,** 311.
Purnell, H. (1962). "Gas Chromatography." Wiley, New York and London.
Rohrschneider, L. (1966). *J. Chromatog.* **22,** 6.

Chapter 2

Derivatization

I. Introduction

A large number of drugs are too polar to be chromatographed directly, but they can be modified often by chemical or physical means to yield less polar compounds. This process is called derivatization. By careful choice of derivatization procedure, the original drug can be converted to a more volatile compound. Decreased polarity and increased volatility reduce adsorption on the stationary phase and result in more symmetric peaks and lower limits of detectability. Formation of a derivative that is less polar than the drug itself allows the column to be operated at a lower temperature, thus prolonging its life, and reducing the possibility of thermal decomposition of the drug.

There are many derivatizing reagents available, and they can be conveniently arranged in three groups: (i) silylating reagents, (ii) nonhalogenating reagents, and (iii) halogenating reagents. The silylating reagents are all volatile, reactive organosilicon compounds that readily combine with a range of drugs to yield derivatives that can be detected using flame ionization. Most of these derivatives are, however, easily hydrolyzed to the original drug, and steps must be taken to exclude traces of water from reagents, solvents, and the reaction mixtures. In contrast, the nonhalogenating reagents yield derivatives that are generally very stable and respond well to flame ionization detection. Halogenating reagents are usually employed to produce derivatives with a strong affinity for electrons so that an electron-capture detector can be used. Halogenated derivatives can also be detected using flame ionization, but the sensitivity is lower. Halogenated derivatives are reasonably stable but care should be taken to exclude water from solvents and reagents or else to use an excess of reagent. It is useful to look briefly at the general characteristics of the three types of reagents; later sections of this chapter will deal with the choice of reagent for particular structural groups.

Silylating reagents can be used singly or in combination. One of the earliest reported silylations used a mixture of trimethylchlorosilane (TMCS) and hexamethyldisilazane (HMDS) dissolved in pyridine (Sweeley *et al.*, 1963). Many silylation reactions are complete within minutes of mixing the drug with the silylating mixture, but, if the functional group to be derivatized is sterically hindered, several hours or overnight refluxing may be necessary. Alcoholic hydroxyl groups can be smoothly silylated by a TMCS–HMDS mixture:

$$\mathrm{ROH} \xrightarrow[\mathrm{py}]{\mathrm{TMCS\text{–}HMDS}} \mathrm{ROSi(CH_3)_3}$$

Substituting dimethylchlorosilane for TMCS yields the dimethylsilyl ether $ROSiH(CH_3)_2$, which has a shorter retention time than the trimethylsilyl ether.

Weaker silyl acceptors such as carboxylic acids, phenols, and amides need stronger silylating reagents: bis(trimethylsilyl)acetamide (BSA) or bis(trimethylsilyl)trifluoroacetamide (BSTFA) are the most frequently used. A typical reaction for BSA would be

$$\mathrm{RCO_2H} + \underset{\mathrm{(BSA)}}{\mathrm{CH_3CON}\!\begin{array}{l}\diagup \mathrm{Si(CH_3)_3}\\ \diagdown \mathrm{Si(CH_3)_3}\end{array}} \longrightarrow \mathrm{RCO_2Si(CH_3)_3} + \mathrm{CH_3CONHSi(CH_3)_3}$$

The mono(trimethylsilyl)acetamide then reacts with another molecule of carboxylic acid, yielding another molecule of the trimethylsilyl ether of the acid and acetamide. BSA is extremely easily hydrolyzed by water, so precautions must be taken to exclude water from the reagent and from any of the solvents used with it such as acetonitrile, pyridine, dimethylformamide, dimethyl sulfoxide, or tetrahydrofuran.

Because of the ready hydrolysis of most silyl derivatives, no attempt is made to purify the product of a silylation reaction; a portion of the reaction mixture is simply injected onto the column of the gas chromatograph. This passage of silylated derivative and excess silylating reagent through the flame ionization detector results in the accumulation of a deposit of silica on the flame tip, and frequent cleaning is necessary. However, by using fluoro analogs such as BSTFA, hydrogen fluoride is produced in the flame and carries off the deposited silica. The two by-products of silylation with

BSTFA, monotrimethylsilyltrifluoroacetamide and trifluoroacetamide, are very volatile, and this reagent could be used in place of BSA where the GC peaks of the by-products are found to interfere with the peak of the silylated drug. The most volatile trimethylsilylacetamide available is *N*-methyl-*N*-trimethylsilyltrifluoroacetamide (MSTFA) with a donor strength similar to that of BSTFA (Donike, 1969); the reaction by-product *N*-methyltrifluoroacetamide is more volatile than MSTFA itself.

The silylation of very sensitive compounds can be carried out under mild conditions using trimethylsilyl imidazole, TMSIm:

$$ROH + (CH_3)_3SiN\text{(imidazole)} \longrightarrow ROSi(CH_3)_3 + HN\text{(imidazole)}$$

TMSIm

The reaction products are the silylated derivative and the weak base imidazole. Moderately hindered groups, for example, the hydroxyl group in 11 β-hydroxysteroids, can be silylated by a mixture of BSA and TMCS (5:1) (Chambaz and Horning, 1969) while even the most hindered 17α-hydroxysteroids will yield to TMSIm–BSA–TMCS (3:3:2) (Sakauchi and Horning, 1971). The main disadvantage associated with silylating reagents is that the derivatives cannot be separated from the reaction mixture (e.g., by extraction or thin-layer chromatography) for fear of their decomposition. This means that cocktails of highly reactive compounds have to be injected onto the GC column, and this is not possible in the case of stationary phases containing hydroxyl groups (Carbowax, PEG, etc.) without radically altering the characteristics of the phase.

Osman and Hill (1982) have described their use of a silicon selective detector, which they claim to be more reliable than either flame ionization or thermal conductivity. A standard FID system was converted to a silicon selective device by interchanging the oxygen and hydrogen inlets and locating the collector electrode a distance of approximately 10 cm above the flame.

Nonhalogenating reagents, in contrast, produce derivatives that can be purified by solvent extraction, thin-layer chromatography, and other techniques. It is not necessary to inject highly reactive mixtures onto the column, the life of the column is prolonged, and detector contamination is greatly reduced. Derivatization with nonhalogenating reagents yields compounds that can be detected by flame ionization. Electron capture can only be used when the original drug contains a group with an affinity for electrons. Many drugs contain functional groups with replaceable hydrogen atoms, e.g., carboxylic acids, phenols, and amines, and these can all be rendered less polar by derivatization: acids to esters, phenols to acetates, amines to amides, and quaternary ammonium compounds to tertiary amines. A wide range of nonhalogenating reagents is available, the most commonly used being the lower anhydrides (acetic, propionic, and butyric), acyl chlorides, diazoalkanes, dimethyl sulfate, and alkyl iodides. The choice of reagent is governed by the functional group to be derivatized, and examples will be given in the following sections. Derivatization is usually carried out by dissolving the drug in a suitable solvent, adding the reagent and, if necessary, a catalyst and heating until derivatization is complete; a number of reagents react rapidly at room temperature. Many nonhalogenated derivatives are stable to water.

Halogenating reagents are usually used to prepare derivatives that respond to electron-capture detection. The halogenated analogs such as chloroacetic, trifluoroacetic, pentafluoropropionic, and heptafluorobutyric anhydrides are used to derivatize phenols, alcohols, and amides, while pentafluorobenzyl bromide and 2,4-dinitrofluorobenzene can be used on any compound with an easily replaceable hydrogen atom. Halogenated derivatives are generally stable to water with the exception of the trifluoroacetates of phenols. The by-products of many derivatizing reactions involving halogenating reagents are much more polar than those produced in reactions using the nonhalogenating analogs, e.g., in acylations using trifluoroacetic anhydride, trifluoroacetic acid is produced (pK_a 0.3) while a similar derivatization with acetic anhydride produces acetic acid (pK_a 4.8). Care must be taken since strongly acidic by-products can lead to undesirable side reactions (see page 44). Derivatization under milder conditions is possible, using reagents such as trifluoroimidazole.

Silylating reagents with halogen atoms are also available, and these can be used to prepare derivatives with greater electron affinity. The most commonly used are 1,3-bis(chloromethyl)-1,1,3,3-tetramethyldisilazane (CMTMDS), chloromethyldimethylchlorosilane (CMDMCS), and bromomethyldimethylchlorosilane (BMDMCS). CMDMCS yields derivatives with retention times longer than those of the corresponding trimethylsilyl compounds but shorter than those of the corresponding chloroacetate. BMDMCS derivatives are more strongly electron capturing than the chloroacetates and approach the heptafluorobutyrates in sensitivity; their retention times, however, are approximately five times greater than those of the corresponding trimethylsilyl derivatives.

Derivatization can be carried out in a test tube and the extract purified before injection onto the column of the gas chromatograph; an alternative technique is "on column" derivatization. This involves injecting the drug onto the column followed by the derivatizing reagent. The reagent is chosen so that it is more volatile than the drug. This is usually a simple matter since most commonly used derivatizing reagents are very volatile. The reagent overtakes the drug and the derivative is formed on the column of the chromatograph or in the heated injection port and then passes through the column to be detected in the normal way. Examples of this type of derivatization are given on page 41.

The rest of this chapter is devoted to the reactions used to derivatize particular functional groups of drugs and the final section deals with the influence of reaction conditions on derivatization.

II. Acids

A. Carboxylic Acids

1. Silylating Reactions

A number of reagents can be used to convert carboxylic acids to their volatile trimethylsilyl ethers. Fenoprofen, the anti-inflammatory agent, has been measured in plasma after derivatization with hexamethyldisilazane in carbon disulfide (Nash *et al.*, 1971):

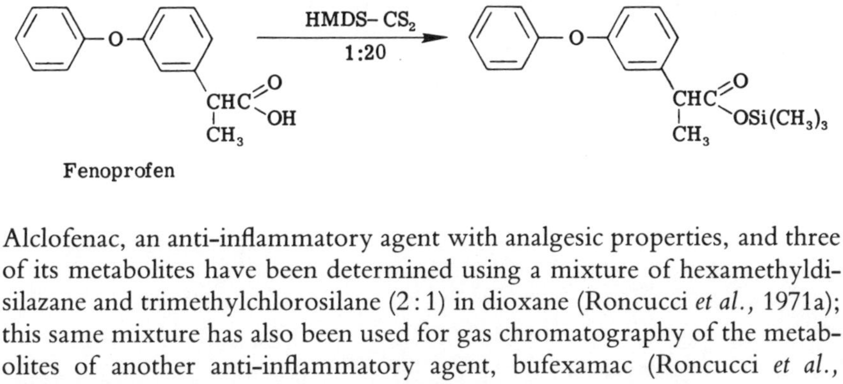

HMDS– CS_2 1:20

Fenoprofen

Alclofenac, an anti-inflammatory agent with analgesic properties, and three of its metabolites have been determined using a mixture of hexamethyldisilazane and trimethylchlorosilane (2 : 1) in dioxane (Roncucci *et al.*, 1971a); this same mixture has also been used for gas chromatography of the metabolites of another anti-inflammatory agent, bufexamac (Roncucci *et al.*, 1971b). Mycophenolic acid, an antibiotic isolated from *Penicillium* cultures, has been converted to the trimethylsilyl ether using trimethylsilylimidazole in carbon disulfide (Bopp *et al.*, 1972), while mycotoxin penicillic acids have been silylated by a mixture of bis(trimethylsilyl)acetamide and trimethylchlorosilane (Pero *et al.*, 1972). Acetylsalicylic acid has been successfully measured in plasma and urine following derivatization with bis(trimethylsilyl)trifluoroacetamide (Thomas *et al.*, 1973).

2. *Nonhalogenating Reactions*

The reagents most commonly used to esterify carboxylic acids are methanolic hydrogen chloride, methanol–boron trifluoride, methanol–boron trichloride, diazoalkanes, dimethyl sulfate, and methyl iodide. Methanolic hydrogen chloride can be prepared by passing hydrogen chloride gas through anhydrous methanol, but a more convenient method uses the reaction between acetyl chloride and dry methanol:

$$CH_3COCl + CH_3OH \rightleftharpoons CH_3CO_2CH_3 + HCl$$

A few drops of acetyl chloride added with great care to 10 ml of methanol can be used immediately to prepare the methyl ester of a carboxylic acid; by substituting ethanol for methanol, the ethyl ester can be obtained. Addition of a few drops of concentrated sulfuric acid to methanol also produces a potent esterifying reagent. The advantage of methanolic hydrogen chloride is that, after derivatization is complete, the excess reagent can be easily removed by evaporation. Higher esters can also be prepared by transesterification, refluxing the methyl ester with an excess of ethanol, propanol, butanol, etc. As mentioned above, methylation can be carried out using boron trichloride–methanol, and this combination is more stable than boron trifluoride–methanol.

The anti-inflammatory drug indomethacin has been measured in plasma following esterification with diazoethane (Ferry *et al.*, 1974):

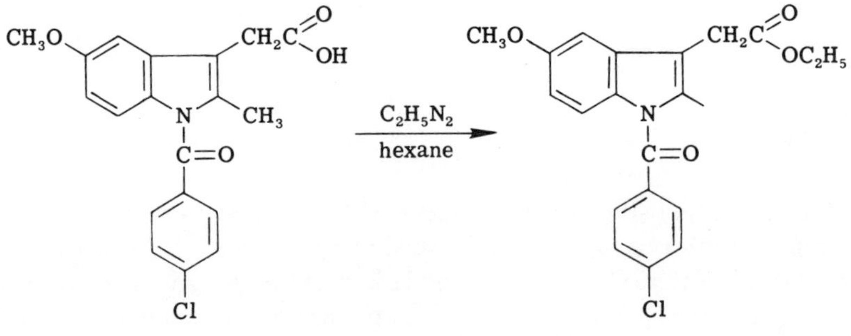

CH_3O, $CH_2C(=O)OH$, N, CH_3, C=O, Cl — Indomethacin $\xrightarrow[\text{hexane}]{C_2H_5N_2}$ CH_3O, $CH_2C(=O)OC_2H_5$, N, C=O, Cl

Indomethacin

Diazoethane was chosen rather than the more commonly used diazomethane in order to allow the determination of indomethacin in the presence of a major metabolite, desmethylindomethacin; the esterification with diazomethane would have produced the *same* derivative from both the original drug and its metabolite. Probenecid has been measured in plasma following esterification with dimethyl sulfate (Sabih *et al.*, 1971), and the determination of traces of salicylic acid in aspirin tablets has been reported (Laik Ali,

1973). The latter method uses methyl iodide and potassium carbonate to esterify the salicylic acid, but a column chromatography step is necessary, before derivatization, to separate salicylic acid from acetylsalicylic acid because the potassium carbonate could cause some hydrolysis of acetylsalicylic acid.

Ibuprofen, an anti-inflammatory, analgesic, and antipyretic agent, has been rapidly esterified using 1,1′-carbonyldiimidazole (Kaiser and Vangiessen, 1974).

Ibuprofen (RCO_2H) + 1,1′-Carbonyldiimidazole → RCOIm

then, $RCOIm \xrightarrow[\text{triethylamine}]{CH_3OH} RCO_2CH_3 + ImH$

The imidazolide formation was completed within 1 min, and conversion to the methyl ester was almost instantaneous. A number of other less frequently used derivatives of carboxylic acids have also been described: *p*-benzyl esters (Watson and Crescuolo, 1970), toluidines (Umeh, 1970a), anilides (Umeh, 1970b), *p*-bromophenacyl, and *p*-phenylphenacyl (Umeh, 1971).

3. Halogenating Reactions

Halogenated silylating agents yield halogenated derivatives with carboxylic acids: the prostaglandin PGF has been measured as the bromosilyl methyl ester (Jouvenaz *et al.*, 1973). Esterification with boron trichloride–2-chloroethanol will produce a derivative with a high affinity for electrons, while pentafluorobenzyl bromide forms a strongly electron-capturing ester. This latter reagent has been used in the gas chromatographic determination of flurbiprofen in plasma (Kaiser *et al.*, 1974):

Flurbiprofen $\xrightarrow[K_2CO_3]{BrH_2C\text{-}C_6F_5}$ pentafluorobenzyl ester

B. Barbituric Acids

Although electron capture has been reported as being more sensitive than flame ionization for the detection of barbiturates (Gudzinowicz and Clark, 1965), almost all the work to date has been carried out using flame ionization. Barbiturates can be measured as the free acids (Inaba and Kalow, 1972) and even as the sodium salts (Welton, 1970), if formic acid is added to the carrier gas. One author has recommended that they be measured quantitatively without derivatization while carrying out qualitative identification using the trimethylsilyl ethers (Street, 1971). The same author stresses the need for treating the diatomaceous earth support with the decomposition products of a silicone gum rubber such as SE-30 or SE-52 at 400°C in order to reduce adsorption. The addition of formic acid to the carrier gas has been reported to lead to increased sensitivity using flame ionization (Welton, 1970).

1. Silylating Reactions

Barbital, Dial, butethal, amobarbital, and pentobarbital have been chromatographed as trimethylsilyl ethers after derivatization with bis(trimethylsilyl)acetamide (Street, 1971). One publication reports that the trimethylsilyl derivative of phenobarbital is stable for only 15 min (Kupferberg, 1972). Silylation is rarely used, most workers preferring to chromatograph the free acids or the nonhalogenated derivatives.

2. Nonhalogenating Reactions

A wide choice of reagents is available to alkylate barbiturates. Diazomethane has been used to methylate 5-(2,3-dihydroxypropyl)-5-neopentylbarbituric acid, the principle metabolite of nealbarbital (Gilbert *et al.*, 1974), and dimethyl sulfate has been used to methylate barbiturates present at low concentrations in biological fluids (Stewart *et al.*, 1969). Methylation can also be carried out using bases like tetramethylammonium hydroxide and phenyltrimethylammonium hydroxide. In a study of alkylation as a method for derivatization of barbiturates (Greeley, 1974), tetramethylammonium hydroxide was reported to be the method of choice for methylation since its stability was greater than that of phenyltrimethylammonium hydroxide, and a higher concentration of the reagent was available commercially. Good separations were obtained with the methylated derivatives, but some barbiturates in use are already methyl derivatives of other barbiturates, e.g., mephobarbital is the methyl derivative of phenobarbital. The *n*-butyl derivatives were better than either the methyl or ethyl derivatives, giving a linear flame ionization response over the range 0.02 to 4.00 μg.

The extractive alkylation of barbiturates has been described (Ehrsson, 1974). The barbiturate forms an ion pair in aqueous solution with a quaternary ammonium salt, which is extracted into an organic phase containing the alkylating agent, e.g., methyl iodide. A microscale procedure for the benzylation of small samples (5 μl) of phenobarbital solution has been described (Dünges, 1973). The reaction is carried out in acetone in the presence of potassium carbonate. One disadvantage of forming a high molecular weight derivative such as the N,N'-dibenzyl is that poor resolution of closely related compounds results. On a polar column of 3% OV-225 programmed from 130 to 220°C at 6°C per minute, barbital had a retention time of approximately 29 min as the N,N'-dibenzyl, while the corresponding derivative of phenobarbital had a retention time of 31 min. The methoxy derivatives of barbiturates can be formed smoothly, using N,N-dimethylformamide (Venturella *et al.*, 1973):

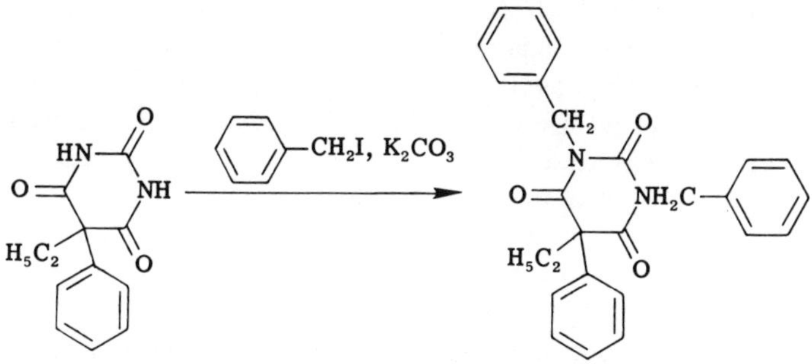

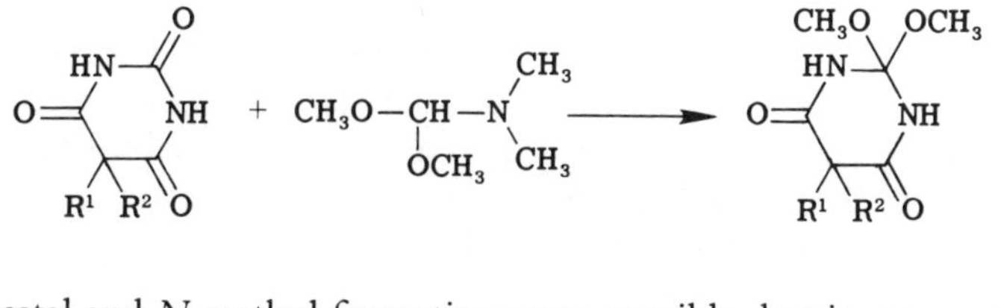

Both acetal and N-methyl formation seem possible, but it appears that the polarization of the C-2 carbonyl takes place more readily than proton abstraction, hence the methoxy rather than the N-methyl derivative is formed.

3. Halogenating Reactions

Halogenating reactions have been described for the barbiturates. Phenobarbital, mephobarbital, barbital, and hexobarbital have all been converted to pentafluorobenzyl derivatives, using pentafluorobenzyl bromide with triethylamine as a catalyst (Walle, 1975). The use of a catalyst is critical since no derivatization takes place in its absence; at low concentrations of catalyst, barbital gave a mixture of mono- and dipentafluorobenzyl derivatives. Potassium carbonate, a useful catalyst in the alkylation of carboxylic acids, was reported to cause hydrolysis of the barbiturates. Interestingly, the benzylation of phenobarbital, described in the section on nonhalogenating reactions, does not mention hydrolysis even though the derivatization was carried out in the presence of potassium carbonate. Phenobarbital has been measured in saliva after conversion to the bispentafluorobenzyl derivative using pentafluorobenzyl bromide in methylene chloride (Gyllenhaal *et al.*, 1976); no catalyst was used.

C. Amino Acids

Since amino acids contain two polar groups, it is necessary to derivatize both in order to measure low concentrations by gas chromatography.

1. Silylating Reactions

One of the earliest reports suggested a one-step reaction, using a mixture of HMDS and TMCS (Ruhlmann and Giesecke, 1961), while trimethylsilyldiethylamine (Smith and Sheppard, 1965) and *N*-trimethylsilyl-*N*-methylacetamide (Birkofer and Donike, 1967) have also been proposed. A comparison of reagents has been published (Smith and Shewbart, 1969) and includes trimethylsilyldimethylamine, trimethylsilyldiethylamine, BSA, BSTFA, *N*-trimethylsilyl-*N*-methylacetamide, and trimethylsilylimidazole. The authors concluded that the trimethylsilylamines were the most stable and resulted in derivatives more volatile than the trimethylsilylamides. A detailed evaluation of a single-step derivatization of the 20 protein amino acids using BSTFA in acetonitrile, has been published (Gehrke and Leimer, 1971). The extracts containing the silyl derivatives were stable for up to 8 days, provided that traces of water were excluded.

2. Nonhalogenating Reactions

Amino acids can be chromatographed successfully as *N*-acyl alkyl esters: the carboxyl group is derivatized first with alcoholic HCl, followed by conversion of the amine to an amide with an organic anhydride. A study of 24 amino acids as *N*-butyryl *n*-propyl esters has been published (Gehrke and Takeda, 1973). Derivatization of both acid and amine functions with a single reagent has also been described (Pettit and Stouffer, 1970; Blessington and Fiagbe, 1972):

$$\mathrm{R{-}CH(NH_2){-}C(=O)OH} \xrightarrow[\text{iodide}]{\text{isopropyl}} \mathrm{R{-}CH(NH{-}CH(CH_3)_2){-}C(=O){-}O{-}CH(CH_3)_2}$$

The yield varied considerably depending on the amino acid. Volatile derivatives can be formed under mild conditions by the addition of benzaldehyde or 2,4-pentanedione to the methyl esters of amino acids (Mitchell, 1973):

$$\mathrm{H_3C{-}C(OH){=}CH{-}C(=O){-}CH_3} + \mathrm{H_2NCH(R)CO_2CH_3} \longrightarrow \mathrm{H_3C{-}C(N(H)CH(R)CO_2CH_3){=}CH{-}C(=O){-}CH_3}$$

3. *Halogenating Reactions*

The most sensitive technique for the detection and estimation of amino acids is gas chromatography of derivatives with good electron-capturing properties. The first report (Zomzely *et al.*, 1962) of such work described the formation of the trifluoroacetamide of the amine group, followed by conversion of the carboxyl group to the butyl ester; later workers carry out esterification first. This approach allows the detection of picogram amounts (Zumwalt *et al.*, 1971), and the 20 protein amino acids can be separated on a single column (Gehrke *et al.*, 1971). A simple apparatus for converting amino acids to their *N*-trifluoroacetyl *n*-butyl esters in a single vial has been described (Mee and Brooks, 1971).

A comparison of the gas chromatograph and the autoanalyzer for the determination of amino acids has been made (Pellizari *et al.*, 1971), and the accuracy of GLC was found to be lower. It remains, however, a very attractive method where limited quantities of sample are available, e.g., in clinical studies involving children. A good overall review of amino acid derivatization has been written by Hušek and Macek (1975).

III. Bases

A. Primary and Secondary Amines

Although many amines are volatile enough to be chromatographed directly, derivatization is usually carried out to increase volatility, decrease peak tailing, and allow the detection of much lower quantities.

1. *Silylating Reactions*

Amines are now rarely derivatized using silylating reagents, but a number of biologically important amines have been chromatographed as trimethylsilyl ethers (Capella and Horning, 1966), and silylation has been used to prepare some aminoglycosidic antibiotics for GLC (Omoto *et al.*, 1971).

2. *Nonhalogenating Reactions*

Amines can be readily converted to amides, using acid anhydrides or acyl chlorides. For example, the antidepressant dibenzepine and its five metabolites have been measured in urine following derivatization with a mixture of acetic anhydride and pyridine (1 : 1) (De Leenheer and Heyndrickx, 1973). The anticonvulsant mexiletine has been determined in body fluids after derivatization with butyric anhydride (Kelly *et al.*, 1973):

$$2,6\text{-}(CH_3)_2C_6H_3\text{-}OCH_2CH(CH_3)NH_2 \xrightarrow[\text{anhydride}]{\text{butyric}} 2,6\text{-}(CH_3)_2C_6H_3\text{-}OCH_2CH(CH_3)NHC(=O)CH_2CH_2CH_3$$

Mexiletine

The butyryl derivative was chosen, rather than the acetyl or propionyl ones, since it gave a better separation from lignocaine, one of the drugs likely to be administered with mexiletine.

The reaction of primary amines with simple ketones to yield Schiff bases has been used to help identify these amines: the difference in retention time

$$C_6H_5\text{-}CH_2CH(CH_3)NH_2 + O{=}C(CH_3)_2 \longrightarrow C_6H_5\text{-}CH_2CH(CH_3)N{=}C(CH_3)_2$$

Amphetamine Acetone Schiff base

between the free amine and the Schiff base is characteristic. This procedure has been applied to amphetamine (Beckett and Rowland, 1964).

3. Halogenating Reactions

By far the most frequently used derivatives for the analysis of low concentrations of amines are the halogenated amides. These compounds, which have excellent electron-capturing properties, are formed by the action of the appropriate anhydride on the primary or secondary amine. For example, amphetamine, methamphetamine, fenfluramine, and chlorphentermine have all been measured at nanogram per milliliter concentrations in blood after derivatizing with heptafluorobutyric anhydride (Bruce and Maynard, 1969). The trifluoroacetamide derivative is also commonly used, and successful gas chromatography has been carried out using the amides formed with mono- and trichloroacetic anhydride, chloroacetyl chloride, pentafluorobenzoyl chloride, and 2,4-dinitrofluorobenzene. Detailed studies have been published comparing the electron-capture sensitivity of the different derivatives (Matin and Rowland, 1972; Moffat *et al.*, 1972). For secondary amines, the best reagent appears to be pentafluorobenzoyl chloride, while for primary amines the choice is between this reagent and pentafluorobenzaldehyde. The presence of a polarizable C=O or C=N adjacent to an aromatic ring appears to enhance the electron-capturing effect, while reduction of the C=O group markedly lowers the response.

Derivatization under mild conditions can be performed, using trifluoroimidazole or *N*-methylbistrifluoroacetamide:

$$RNH_2 + \underset{N\text{-methylbistrifluoroacetamide}}{CF_3\overset{O}{\overset{\|}{C}}-\overset{CH_3}{\overset{|}{N}}-\overset{O}{\overset{\|}{C}}-CF_3} \longrightarrow \underset{\mathbf{I}}{RNH-\overset{O}{\overset{\|}{C}}-CF_3} + \underset{\mathbf{II}}{CH_3NH-\overset{O}{\overset{\|}{C}}-CF_3}$$

The products of the above reaction are both neutral amides, and **II** is very volatile and easily removed by evaporation of the reaction mixture (Donike, 1973).

The separation of optically active amines can be achieved by derivatizing with *N*-trifluoroacetyl-L-prolyl chloride (TPC). The separation of the optical isomers of norephedrine, norpseudoephedrine, ephedrine, *N*-ethylnorephedrine, and *N*-ethylnorpseudoephedrine has been reported (Beckett and Testa, 1973).

The halogenated reagents used to derivatize amines are highly reactive, and unexpected products can occur: the reaction of desipramine with trifluoroacetic anhydride yielded a product that was shown to have two oxygen and six fluorine atoms. Mass spectrometric studies showed it to be formed as follows (Walle *et al.*, 1972):

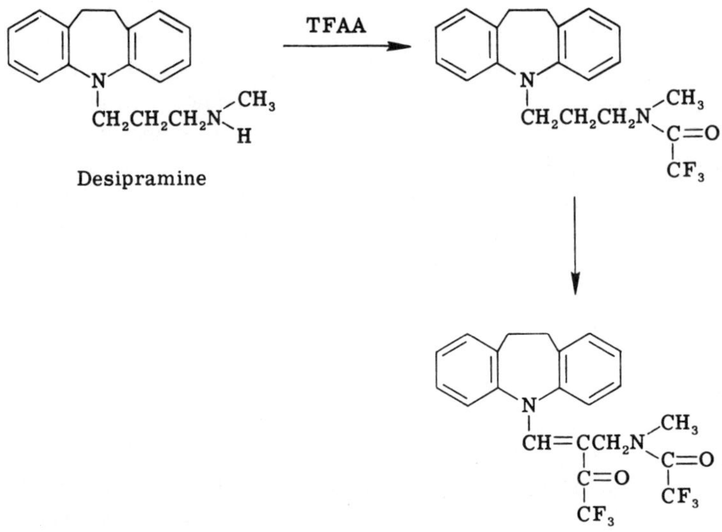

TFAA

$CH_2CH_2CH_2N(CH_3)H$

Desipramine

$CH_2CH_2CH_2N(CH_3)C{=}O(CF_3)$

$CH{=}C(C{=}O\,CF_3)CH_2N(CH_3)C{=}O(CF_3)$

B. Tertiary Amines

A number of drugs containing the tertiary amine group have been chromatographed directly. The analgesic propoxyphene has been measured in the nanogram range in blood (Verebely and Inturrisi, 1973), while the sedative GP 41299 has been studied using the nitrogen-selective flame ionization detector (Besserer *et al.*, 1972). Derivatization with silylating reagents is not possible since no "active" hydrogen atoms are attached to the amine group.

1. Nonhalogenating Reactions

Tertiary amines have been estimated indirectly by first exhaustively methylating, then degrading thermally and measuring the resulting olefin (Hucker and Miller, 1968). This approach has been used to characterize tertiary amines (Street, 1972). After extraction of blood or liver, methylation was carried out using iodomethane in ether for several minutes in a boiling-water bath. The volatile components in the reaction mixture were then evaporated, the residue was redissolved in methanol, and a knife point of silver oxide was added. The mixture was then allowed to stand for 1 min. A portion of the mixture was then injected onto an SE-30 column at 200°C with the injection port heated to 300°C. The chromatograms obtained using a range of tertiary amines were found to be of three types:

1. a peak with a retention time the same as the parent drug, i.e., when the nitrogen atom surrounded by groups that prevent quaternization, such as benzphetamine:

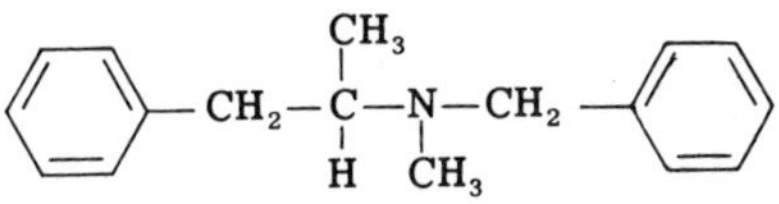

$$C_6H_5-CH_2-CH(CH_3)-N(CH_3)-CH_2-C_6H_5$$

2. a peak with a retention time shorter than the parent drug, i.e., thermal degradation of the quaternary compound has taken place; amitriptyline behaves in this way:

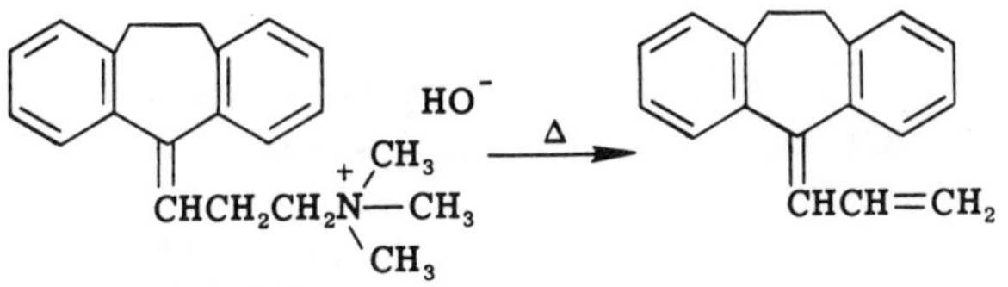

$$=CHCH_2CH_2\overset{+}{N}(CH_3)_3\ HO^- \xrightarrow{\Delta} =CHCH=CH_2$$

Amitriptyline

3. two peaks each with a retention time shorter than the parent but close to each other, i.e., thermal degradation has taken place in two different ways; some phenothiazines produce this type of degradation:

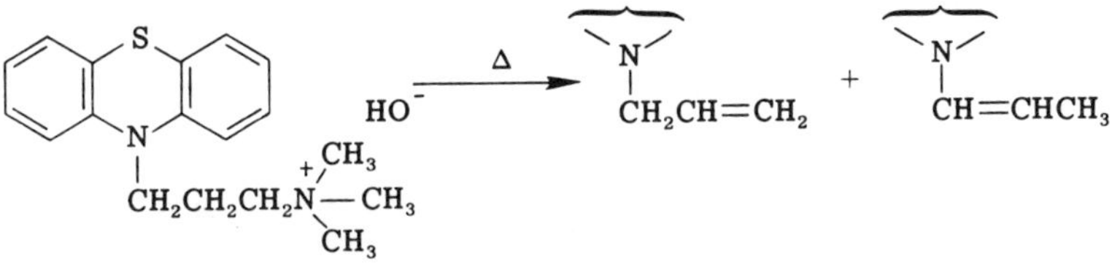

$$N-CH_2CH_2CH_2\overset{+}{N}(CH_3)_3\ HO^- \xrightarrow{\Delta} N-CH_2CH=CH_2 + N-CH=CHCH_3$$

2. Halogenating Reactions

In order to provide a derivative that is electron capturing, the amine can be demethylated to the secondary amine, which can then be converted to an amide in the normal way. Demethylation is carried out by formation of a carbamate, using chloroformate, which is then decomposed by acid to give the secondary amine. The complete reaction sequence is shown below.

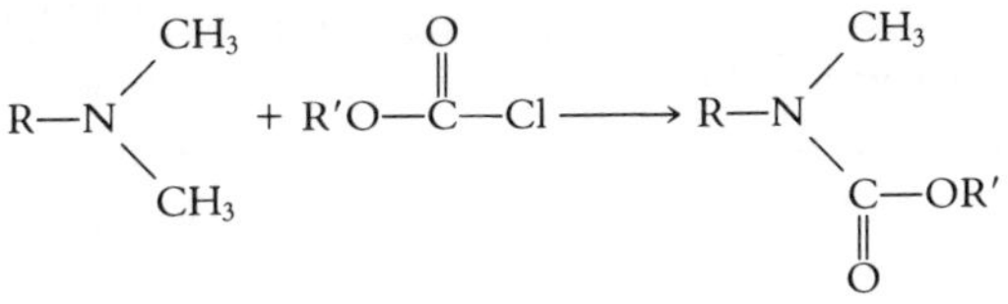

$$R-N(CH_3)_2 + R'O-\overset{O}{\overset{\|}{C}}-Cl \longrightarrow R-N(CH_3)-\underset{O}{\underset{\|}{C}}-OR' \quad (1)$$

$$R{-}N(CH_3){-}C({=}O){-}OR' + H^+ \longrightarrow R{-}N(CH_3){-}H \qquad (2)$$

$$R{-}N(CH_3){-}H + (CF_3CO)_2O \longrightarrow R{-}N(CH_3){-}C({=}O){-}CF_3 \qquad (3)$$

A study of the demethylation of tertiary amines has shown that methyl chloroformate is the most suitable reagent (Hartvig and Vessman, 1974). An interesting variation of this reaction was demonstrated in a method developed to measure the hypnotic *N,N*-dimethyldibenzo[*b,f*]thiepin-10-methylamine. Reaction with ethyl chloroformate gave, instead of a secondary amine, a chloromethylene derivative with good electron-capturing properties (Degen and Riess, 1977).

$$\text{GP 41299 } (H_3C{-}N(CH_3){-}CH_2{-}\text{dibenzothiepin}) \xrightarrow[\text{diisopropyl-ethylamine}]{C_2H_5OC({=}O)Cl} ClCH_2{-}\text{dibenzothiepin}$$

C. Quaternary Ammonium and Guanidine Compounds

A great deal of ingenuity has been used to devise methods of adapting the highly polar quaternary ammonium compounds for gas chromatography. Direct injection of quaternary ammonium hydroxides onto alkaline columns results in the corresponding tertiary amines and olefins (Barry and Saunder, 1971). Acetylcholine has been reduced with potassium borohydride, and the ethanol produced has been measured on a Carbowax 6000 column (Stavinoha and Ryan, 1965). The most widely used method appears to be conversion to the corresponding tertiary amine by either pyrolysis

(Green and Szilagyi, 1974) or reaction with sodium benzenethiolate (Shamma *et al.*, 1966):

$$RCH_2CH_2\overset{+}{N}(CH_3)_3\ X^- \xrightarrow{550^\circ C} RCH_2CH_2N(CH_3)_2 + CH_3X$$

$$RCH_2CH_2\overset{+}{N}(CH_3)_3\ X^- \xrightarrow{^-S-C_6H_5} RCH_2CH_2N(CH_3)_2 + CH_3S-C_6H_5$$

Using mass fragmentography, 10 pmol of acetylcholine can be detected (Hasegawa *et al.*, 1982).

Fewer reports have been published on the gas chromatography of the very basic guanidine compounds. The antihypertensive drug guanoxan has been chromatographed after acetylation with a mixture of acetic anhydride and pyridine (Jack *et al.*, 1972), and the hypoglycemic biguanides buformin and phenformin have been converted to *s*-triazines on column (Wickramasinghe and Shaw, 1972). A study of several guanidine drugs, including guanethidine and guanoxan, described the conversion of the guanidines to the corresponding amines by heating in strongly alkaline solution (Hengstmann *et al.*, 1974):

$$\underset{\text{Guanethidine}}{(CH_2)_7N-CH_2CH_2NHC(=NH)NH_2} \xrightarrow[110^\circ C]{40\%\ KOH} (CH_2)_7N-CH_2CH_2NH_2$$

The amines can then be further derivatized, e.g., using trifluoroacetic anhydride.

Erdtmansky and Goehl (1975) have derivatized the antihypertensive drug debrisoquine and its hydroxylated metabolite using hexafluoroacetyl acetone with electron-capture detection. A further conversion step, to the trimethylsilyl ether, was needed for the metabolite. Lennard *et al.* (1977) chromatographed both compounds as their acetylacetone adducts with nitrogen-selective flame ionization detection.

IV. Phenols and Alcohols

A. Silylating Reagents

The silylating reagents most frequently used are BSA, BSTFA, TMCS, HMDS, and TMSIm. The general reaction is

$$ROH + (CH_3)_3SiX \longrightarrow ROSi(CH_3)_3$$

Morphine has been measured in biological fluids after derivatization with a mixture of BSTFA and TMCS (Wilkinson and Leong Way, 1969), and a number of halogenated 8-hydroxyquinolines with antiseptic properties have been chromatographed as trimethylsilyl ethers after reaction with TMSI (Gruber *et al.*, 1972):

R¹ R² OH N TMSIm R¹ R² N O Si(CH$_3$)$_3$

R¹	R²
Cl	Cl
Cl	I
I	I
Cl	H

Other examples of drugs chromatographed as silyl derivatives include neomycin (Vangiessen and Tsuji, 1971), chlorphenesin (Douglas *et al.*, 1970), ethambutol (Richard *et al.*, 1974), vitamin C (Vecchi and Kaiser, 1967), vitamin D (Fisher *et al.*, 1972), and clindamycin. A mixture of TMSIm, BSA, and TMCS (3 : 3 : 2) was used for the silylation of clindamycin-2-palmitate (Brodasky and Sun, 1974).

B. Nonhalogenating Reagents

The nonhalogenating reagents most frequently used are those that convert phenols and alcohols to their corresponding acetates and ethers: anhydrides, acyl chlorides, diazoalkanes, alkyl iodides, and dimethyl sulfate. These derivatives are simple to prepare and are generally very stable. The bactericide iodochlorhydroxyquin has been measured in body fluids as the acetyl (Jack and Riess, 1973) and methoxy derivatives (Degen *et al.*, 1976), while the alkyl esters of *p*-hydroxybenzoic acid have been chromatographed after conversion to the ethers, using diazoalkanes (Wilcox, 1967).

Other reagents have been used for phenols and alcohols: the psychostimulant pipradrol yielded an unexpected product on oxidation with nonaqueous chromic acid (Hartvig, 1974). The product, 6-benzoyl-1,2,3,4-tetrahydropyridine, exhibits strong electron-capturing properties arising from an in-

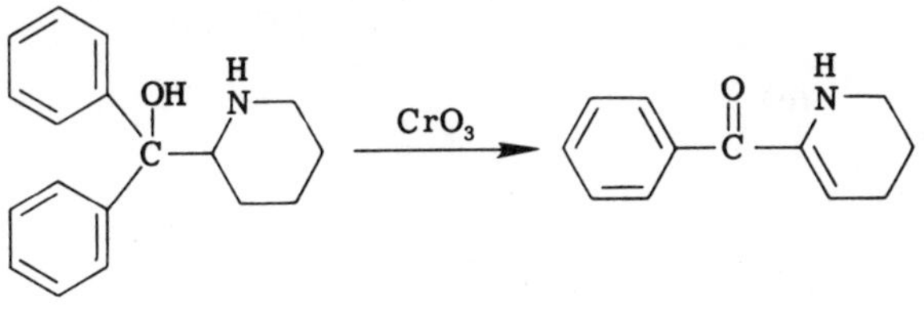

CrO_3

Pipradrol

creased stabilization of a captured electron by intramolecular hydrogen bonding between N—H and C=O. The prostaglandin $F_{2\alpha}$ has been derivatized using *n*-butylboronic acid for the C-9 and C-11 hydroxyl groups, silylation for the C-15 hydroxyl, and esterification for the carboxylic acid function (Kelly, 1972):

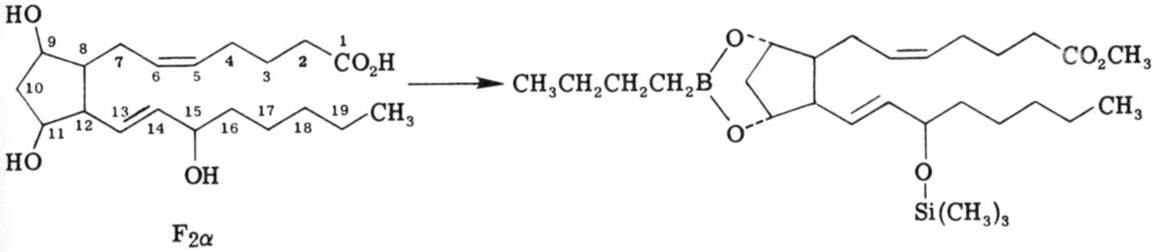

HO HO OH CO_2H CH_3 $CH_3CH_2CH_2CH_2B$ CO_2CH_3 CH_3 O $Si(CH_3)_3$

$F_{2\alpha}$

Boronate derivatives have also been used to characterize catecholamines and related hydroxyamines by GC–MS (Anthony *et al.*, 1970).

C. Halogenating Reagents

Although the trifluoroacetate of an alcohol is generally stable, the pentafluoropropionate, heptafluorobutyrate, or chloroacetate are to be preferred in the case of a phenol because hydrolysis is less likely (McCallum and Armstrong, 1973). The adrenergic β-blocking drugs alprenolol, oxprenolol, and propranolol have all been chromatographed with their alcoholic hydroxyl groups converted to trifluoroacetates (see pages 123–127). Cannabinol has been studied both as the chloroacetate (Schou *et al.*, 1971) and as the pentafluorobenzoate (Garrett and Hunt, 1973):

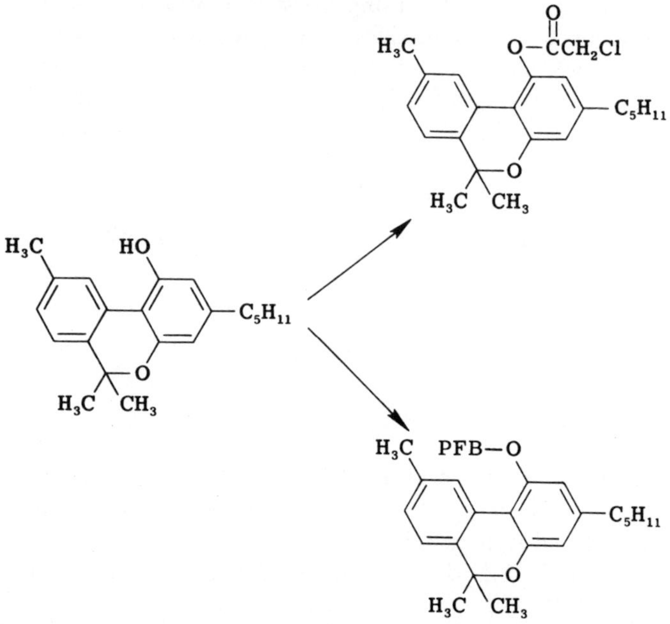

A study of the derivatization of catecholamines has shown that the heptafluorobutyrates give a better response than either the trifluoroacetates or the trimethylsilyl ethers (Sakauchi *et al.*, 1972). The gas chromatography of optically active alcohols has been reported following derivatization with (−)-menthyl chloroformate (Westley and Halpern, 1966).

V. Steroids

Although it is not essential to treat steroids as a separate class since compounds like cholesterol and estrogen can be derivatized by the same reagents used for alcohols and phenols, it is convenient to do so. A number of publications have reviewed derivatization reactions for the GLC of steroids (Kuksis, 1966; Wotiz and Clark, 1966), but only a few selected developments will be considered.

A. Silylating Reagents

All the common silylating reagents have been used to derivatize steroids and include (i) the derivatization of synthetic glucocorticosteroids with a mixture of BSA, TMSIm, and TMCS (Simpson, 1973); (ii) base-catalyzed silylation leading to the preferential formation of enol TMS esters for keto groups without affecting the rate of silylation of hindered hydroxyl groups (Chambaz *et al.*, 1973); and (iii) the use of *tert*-butyldimethylchlorosilane to derivatize hydroxyl groups (Quilliam and Westmore, 1978).

This last reaction is useful in that it introduces a bulky *tert*-butyl group, which then protects the Si—O bond from hydrolysis. This reagent was developed initially to protect the hydroxyl groups of prostaglandins after silylation; the bulky group also offers protection under basic and reducing conditions, and the reaction has been extended to include estrogens, cortisone, and corticosterone.

B. Nonhalogenating Reagents

The most important group of reagents are the boronic acid derivatives that react with corticosteroids to give stable boronates:

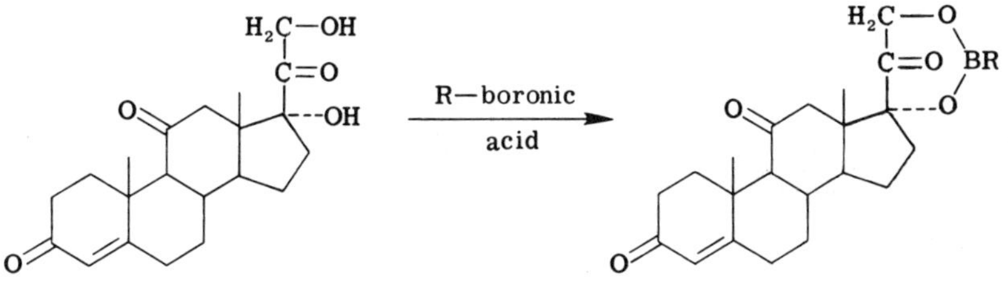

The boronates (R = *n*-butyl, *tert*-butyl, cyclohexyl, and phenyl) have satisfactory GLC properties, although partial decomposition has been reported with the ketol boronates (Brooks and Watson, 1968). Further derivatization is also possible. The acetates of the hydroxysteroid boronates have been formed using acetic anhydride–pyridine, and the keto group of the cyclic boronates of the 17,21-dihydroxyl-20-oxosteroids have been converted to the methoxime. A study of the GC–MS properties of the corticosteroid boronates has been published (Brooks and Harvey, 1971).

C. Halogenating Reagents

The low concentrations of many steroids present in biological fluids are often measured by taking advantage of the sensitivity offered by electron-capture detection: estrogens have been determined as haloacetates (Rajkowski and Broadhead, 1972), heptafluorobutyrates (Wotiz *et al.*, 1969), pentafluorophenylhydrazones (Attal and Eik-Nes, 1968), halomethyldimethylsilyl ethers (Morvay, 1973), and 3,4,5-trihalobenzoyl esters (Morvay, 1973). A study of the TMS, TFA, and HFB derivatives of 21 steroids, including androsterone, 11-hydroxy-, and 11-ketoandrosterone, testosterone, estrone, progesterone, and pregnanediol, on four different stationary phases using FID has been published (Gupta *et al.*, 1971). The authors found that the response factors obtained reflected the sensitivity of the flame ionization detector to the different derivatives. The TMS ethers and HFB derivatives showed greater response factors than the trifluoroacetates, which tended to give multiple peaks. The assay of digoxin in plasma as the digitoxigenin heptafluorobutyrate has been reported, but the overall recovery is only about 25% (Watson and Kalman, 1971). Also, halogenated silylating reagents have been used in the measurement of dehydroepiandrosterone and testosterone (Chapman and Bailey, 1973).

VI. Miscellaneous Derivatization Techniques

A. Ureas

Drugs containing the group R^1R^2-$NCONH_2$ can either be chromatographed directly or be converted to TMS ethers (Kupferberg, 1972) or trifluoroacetamides (Evans, 1974).

B. Phenothiazines

Low concentrations of the phenothiazines promazine, acepromazine, and triflupromazine have been measured in blood by electron-capture detection following bromination (Noonan, 1972). The bromination was found to be very dependent on the substrate concentration; the most suitable brominating mixture was found to be 1% bromine in iodine-saturated cyclohexane.

C. Ring Opening and Closing

A number of reactions have been reported in which ring opening or closing converts a drug to a compound with better chromatographic properties. For example, the benzodiazepines clonazepam and flunitrazepam give only poorly defined peaks when chromatographed directly but can be detected at concentrations of 1 ng/ml in blood when converted to the corresponding benzophenones (DeSilva *et al.*, 1974). An example of ring closing is demonstrated in the method used to measure plasma levels of the

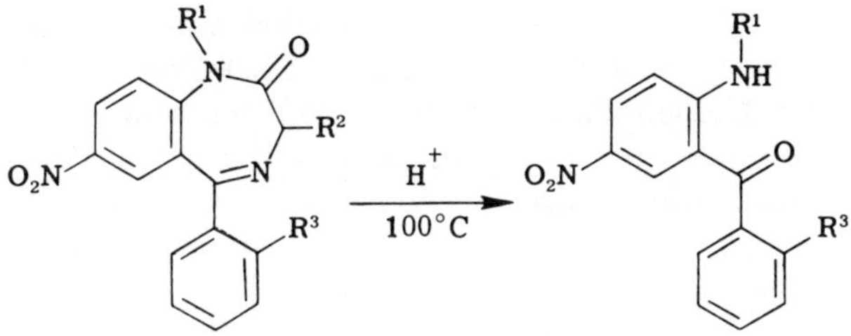

$$\xrightarrow[100^\circ C]{H^+}$$

	R^1	R^2	R^3
Clonazepam	H	H	Cl
Flunitrazepam	CH_3	H	F

antihypertensive hydralazine (Jack *et al.*, 1975). Hydralazine is unstable in aqueous media, and derivatization is carried out by adding the reagents directly to the plasma. Nitrous acid converts hydralazine to tetrazolophthal-

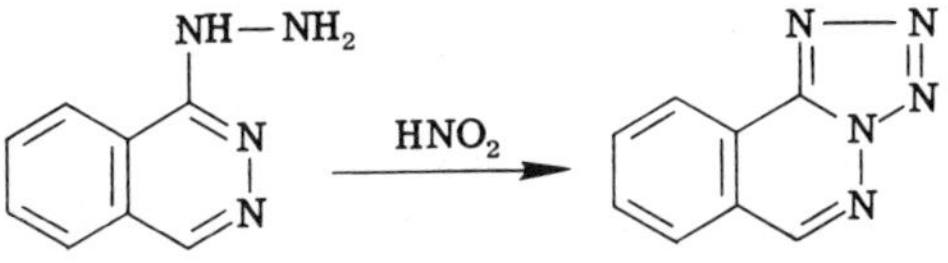

$$\xrightarrow{HNO_2}$$

Hydralazine

Tetrazolo-phthalazine

azine, a compound with a strong affinity for electrons. Using this method, plasma levels of 20 ng/ml could be determined (see also pages 128–130).

D. On-Column Derivatization

It was mentioned briefly in the introduction to this chapter that on-column derivatization is carried out by injecting the compound onto the gas chromatograph and then immediately injecting the derivatizing reagent. Such a tecnique is not to be recommended when electron-capture detection is used because of the possibility of detector contamination, but it can be very useful when flame ionization is used. The drug ethosuximide has been determined after flash heater methylation (Solow and Green, 1971), methimazole has been detected in rat urine as *S*-methylmethimazole following on-column derivatization with methyl iodide in acetone (Stenlake *et al.*, 1970), and tetramethylammonium hydroxide has been used to methylate diphenylhydantoin (Estas and Dumont, 1973).

One major disadvantage of flash heater methylation is that several compounds may give the same derivative, e.g., phenobarbital and 1-methylphenobarbital. Theophylline, theobromine, and their monomethylxanthine metabolites are all converted to caffeine, and to avoid this, tetra-*n*-butylammonium hydroxide has been used (Kowblansky *et al.*, 1973).

VII. The Effect of Reaction Conditions on Derivatization

The factors influencing each derivatization reaction should be investigated so that conditions can be chosen to give the best yield of product. The most important factors to be considered are (i) solvent for the reaction, (ii) temperature, (iii) concentration of reagent, (iv) time of reaction, and (v) use of a catalyst. Many reactions are solvent dependent to a remarkable degree, and it is prudent to try several solvents of differing polarity when investigating the derivatization of a new drug. The trifluoroacetylation of 2-methoxyethylcarbamic acid propyl ester gives very different yields, depending on whether the reaction is carried out in hexane or ethyl acetate (Mellström and Ehrsson, 1974) (Fig. 2.1). Treatment of the antibiotic chloramphenicol with BSA gives the mono-, bis- and trismethylsilyl derivatives, depending on the solvent used (Janssen and Vanderhaeghe, 1973) (Table 2.1). Only by using a mixture of HMDS and TMCS in pyridine could a single derivative, bistrimethylsilylchloramphenicol, be formed.

The temperature at which derivatization is carried out should also be examined, although this is not often as critical a factor as, say, the nature of

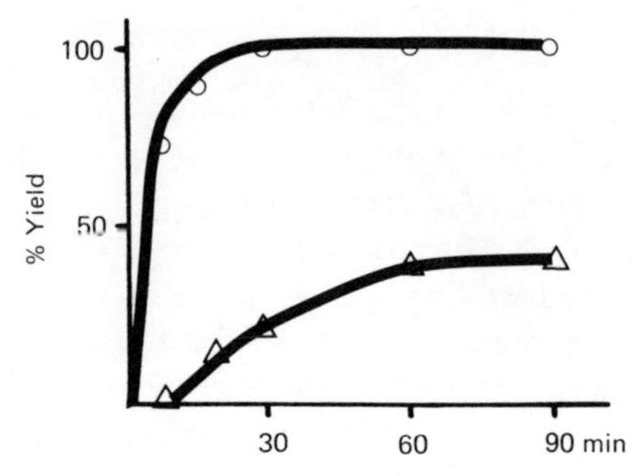

FIG. 2.1. Trifluoroacetylation of 2-methoxyethylcarbamic acid propyl ester in hexane (○) and in ethyl acetate (△). From Mellström and Ehrsson (1974). Adapted with permission from *Acta Pharm. Suecica*.

Table 2.1

Silylation of Chloramphenicol with BSA

Solvent	TMS derivative formed
Acetonitrile	Bis and tris
Pyridine	Bis and tris
Ethyl acetate	Mono, bis, and tris
Chloroform	Mono, bis, and tris

the solvent. Increasing the reaction temperature usually leads to faster reaction times. Care must be taken, however, not to heat the reaction mixture too much since labile substances may decompose or volatile reagents may boil off before the derivatizing reaction takes place.

An important factor that must be controlled and optimized is the concentration of the derivatizing reagent in the reaction mixture. This is illustrated in the method for the determination of theophylline and probenecid in serum samples (Arbin and Edlund, 1974) (Fig. 2.2). The compounds are converted to the butyl esters using dimethylformamide di(*n*-butyl)acetal. As can be seen from Fig. 2.2, the yield falls rapidly if the concentration of the dimethylformamide di(*n*-butyl)acetal in acetone is less than 30%.

Occasionally reaction products can interfere with the starting compound and reduce the final yield. In derivatizing 3-methylindole with trifluoroacetic anhydride (Ehrsson, 1972), it was found that the indole dimer-

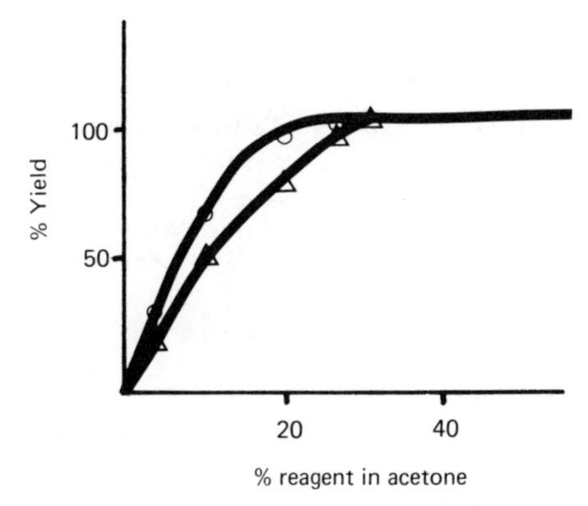

FIG. 2.2. Effect of concentration of reagent on the yields of the butyl esters of theophylline (○) and probenecid (△). From Arbin and Edlund (1974). Adapted with permission from *Acta Pharm. Suecica*.

ized in the presence of the trifluoroacetic acid formed during the reaction, although the *N*-TFA derivative itself was completely stable:

$$\text{3-methylindole (N–H)} + (CF_3CO)_2O \longrightarrow \text{N-}(COCF_3)\text{-3-methylindole} + CF_3CO_2H$$

$$\text{3-methylindole} \xrightarrow{CF_3CO_2H} \text{dimer}$$

The dimerization was overcome by the addition of the base trimethylamine, which neutralized the liberated acid, to the solution of 3-methylindole before trifluoroacetic anhydride was added.

The time needed for reaction should be carefully controlled since it can vary considerably, depending on the compound and the derivative chosen. The rate of imidazolide formation by the anti-inflammatory analgesic ibuprofen is very rapid (Kaiser and Vangiessen, 1974), while the pentafluorobenzylation of flurbiprofen requires about 1 h for the maximum yield to be obtained (Kaiser *et al.*, 1974) (see Fig. 2.3).

Care should also be taken that the reaction is not allowed to continue much beyond the time needed for maximum yield since degradation of the product may take place. This was found to be the case when *gem*-diphenyl substituted compounds were measured after oxidation in alkaline permanganate to benzophenone (Hartvig *et al.*, 1972): after a reaction time greater than 1 h a decrease in yield was found due to degradation of the benzophenone.

The use of a catalyst can also influence the yield. Most simple acetylations are carried out using acetic anhydride with pyridine as the catalyst, but

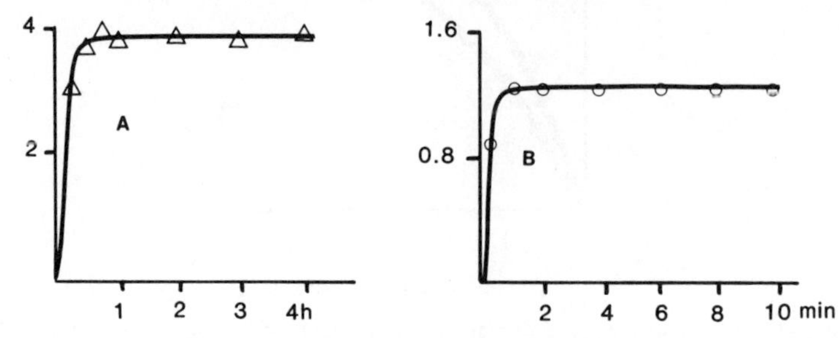

FIG. 2.3. A. Rate of pentafluorobenzylation of flurbiprofen. B. Rate of imidazolide formation by ibuprofen. The *y*-axis measures the derivative peak height ratio (drug : internal standard). From Kaiser *et al.* (1974) and Kaiser and Vangiessen (1974). Adapted with permission of the copyright owner.

Table 2.2

Comparison of Acylation Catalysts

Compound	Catalyst	Yield
Methyl cholate	Pyridine	70% 3,7-Diacetate
	4-Pyrrolidinopyridine	100% 3,7,12-Triacetate
2,4,6-Trimethylphenol[a]	Pyridine	0%
	4-Dimethylaminopyridine	98% Acetate

[a] Structure (2,4,6-trimethylphenol: benzene ring bearing OH, with H_3C and CH_3 at the two ortho positions and CH_3 at the para position)

where steric hindrance is present, pyridine is not effective. Its replacement by a 4-dialkylaminopyridine can spectacularly improve the yield (Höfle and Steglich, 1972) (see Table 2.2).

The acetylation of clindamycin palmitate hydrochloride has been the subject of a kinetic study (Rowe and Machkovech, 1977) (Fig. 2.4). The acetyl-

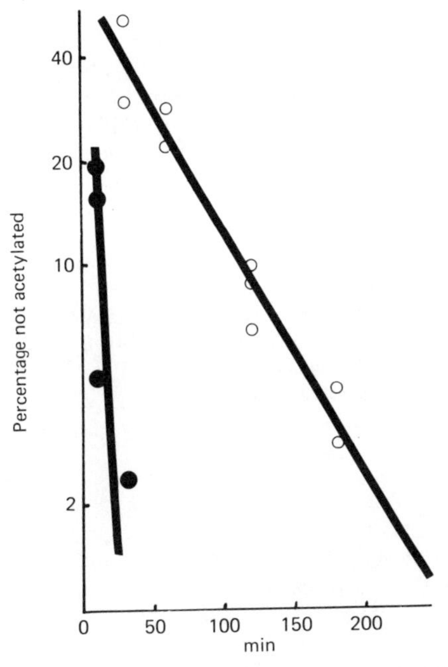

FIG 2.4. Rate of acetylation of clindamycin palmitate hydrochloride with acetic anhydride in pyridine (○) at 95°C and with 4-dimethylaminopyridine (●) added to pyridine at room temperature. Adapted from Rowe and Machkovech (1977) with permission of the copyright owner.

ation reaction time was reduced from 2.5 h at 95°C using pyridine to 30 min at room temperature using pyridine with 4-dimethylaminopyridine. Acylation reactions with halogenated reagents such as trifluoroacetic anhydride and heptafluorobutyric anhydride are often carried out without the use of a catalyst, but caution should be used since, in the case of the drug terodiline, no acylation was found to take place in the absence of the catalyst trimethylamine (Hartvig *et al.*, 1974).

References

Anthony, G. M., Brooks, C. J. W., and Middleditch, B. S. (1970). *J. Pharm. Pharmacol.* **22,** 205.

Arbin, A., and Edlund, P-O. (1974). *Acta Pharm. Suecica* **11,** 249.

Attal, J., and Eik-Nes, K. B. (1968). *Anal. Biochem.* **26,** 398.

Barry, B. W., and Saunders, G. M. (1971). *J. Pharm. Sci.* **60,** 645.

Beckett, A. H., and Rowland, M. (1964). *J. Pharm. Pharmacol.* **16,** 27T.

Beckett, A. H., and Testa, B. (1973). *J. Pharm. Pharmacol.* **25,** 382.

Birkofer, L., and Donike, M. (1967). *J. Chromatogr.* **26,** 270.

Blessington, B., and Fiagbe, N. I. Y. (1972). *J. Chromatogr.* **68,** 259.

Bopp, R. J., and Schirmer, R. E., and Meyers, D. B. (1972). *J. Pharm. Sci.* **61,** 1750.

Brodasky, T. F., and Sun, F. F. (1974). *J. Pharm. Sci.* **63,** 360.

Brooks, C. J. W., and Watson, J. (1969). *In* "Proceedings 7th International Symposium on Gas Chromatography" (C. L. A. Harbourn, ed.), p. 129. Institute of Petroleum, London.

Bruce, R. B., and Maynard, W. R., Jr., (1969). *Anal. Chem.* **41,** 977.

Capella, P., and Horning, E. C. (1966). *Anal. Chem.* **38,** 316.

Chambaz, E. M., and Horning, E. C. (1969). *Anal. Biochem.* **30,** 7.

Chambaz, E. M., Defaye, G., and Madani, C. (1973). *Anal. Chem.* **45,** 1090.

Chapman, J. R., and Bailey, E. (1973). *Anal. Chem.* **45,** 1636.

Degen, P. H., and Riess, W. (1973). *J. Chromatogr.* **85,** 53.

Degen, P. H., Schneider, W., Vuillard, P., Geiger, U. P., and Riess, W. (1976). *J. Chromatogr.* **117,** 407.

De Leenheer, A., and Heyndrickx, A. (1973). *J. Pharm. Sci.* **62,** 31.

De Silva, J. A. F., Puglisi, C. V., and Munno, N. (1974). *J. Pharm. Sci.* **63,** 520.

Donike, M. (1969). *J. Chromatogr.* **42,** 103.

Donike, M. (1973). *J. Chromatogr.* **78,** 273.

Douglas, J. F., Stockage, J. A., and Smith, N. B. (1970). *J. Pharm. Sci.* **59,** 107.

Dünges, W. (1973). *Anal. Chem.* **45,** 963.

Ehrsson, H. (1972). *Acta Pharm. Suecica* **9,** 419.

Ehrsson, H. (1974). *Anal. Chem.* **46,** 922.

Erdtmansky, P., and Goehl, T. J. (1975). *Anal. Chem.* **47,** 750.

Estas, A., and Dumont, P. A. (1973). *J. Chromatogr.* **82,** 307.

Evans, R. T. (1974). *J. Chromatogr.* **88,** 398.

Ferry, D. G., Ferry, D. M., Moller, P. W., and McQueen, E. G. (1974). *J. Chromatogr.* **89,** 110.

Fisher, A. L., Parfitt, A. M., and Lloyd, H. M. (1972). *J. Chromatogr.* **65,** 493.

Garrett, E. R., and Hunt, C. H. (1973). *J. Pharm. Sci.* **62,** 1211.

Gehrke, C. W., and Leimer, K. (1971). *J. Chromatogr.* **57,** 219.

Gehrke, C. W., and Takeda, H. (1973). *J. Chromatogr.* **76,** 63.
Gehrke, C. W., Kuo, K., and Zumwalt, R. W. (1971). *J. Chromatogr.* **57,** 209.
Gilbert, J. N. T., Millard, B. J., Powell, J. W., Whalley, W. B., and Wilkins, B. J. (1974). *J. Pharm. Pharmacol.* **26,** 119.
Greeley, R. H. (1974). *Clin. Chem.* **20,** 192.
Green, J. P., and Szilagyi, P. I. A. (1974). *In* "Choline and Acetylcholine: Handbook of Chemical Assay Methods" (I. Hanin, ed.), p. 151. Raven, New York.
Gruber, M. P., Klein, R. W., Foxx, M. E., and Campisi, J. (1972). *J. Pharm. Sci.* **61,** 1147.
Gudzinowicz, B. J., and Clark, S. J. (1965). *J. Gas Chromatogr.* **3,** 147.
Gupta, D., Breitmaier, E., Jung, G., von Lucadon, G., Pauschmann, H., and Voeller, W. (1971). *Chromatographia* **4,** 572.
Gyllenhaal, O., Brötell, H., and Sandgren, B. (1976). *J. Chromatogr.* **122,** 471.
Hartvig, P. (1974). *Acta Pharm. Suecica* **11,** 109.
Hartvig, P., and Vessman, J. (1974). *Acta Pharm. Suecica* **11,** 115.
Hartvig, P., Sundin, H., and Vessman, J. (1972). *Acta. Pharm. Suecica* **9,** 269.
Hartvig, P., Freij, G., and Vessman, J. (1974). *Acta Pharm. Suecica* **11,** 97.
Hengstmann, J. H., Falkner, F. C., Watson, J. T., and Oates, J. (1974). *Anal. Chem.* **46,** 34.
Höfle, G., and Steglich, W. (1972). *Synthesis,* 619.
Hucker, H. B., and Miller, J. K. (1968). *J. Chromatogr.* **32,** 408.
Hušek, P., and Macek, K. (1975). *J. Chromatogr.* **113,** 139.
Inaba, T., and Kalow, W. (1972). *J. Chromatogr.* **69,** 377.
Jack, D. B., and Riess, W. (1973). *J. Pharm. Sci.* **62,** 1929.
Jack, D. B., Stenlake, J. B., and Templeton, R. (1972). *Xenobiotica* **2,** 35.
Jack, D. B., Brechbühler, S., Degen, P. H., Zbinden, P., and Riess, W. (1975). *J. Chromatogr.* **115,** 87.
Janssen, G., and Vanderhaeghe, H. (1973). *J. Chromatogr.* **82,** 297.
Jouvenaz, G. H., Nugteren, D. H., and Van Dorp, D. A. (1973). *Prostaglandins* **3,** 175.
Kaiser, D. G., and Vangiessen, G. J. (1974). *J. Pharm. Sci.* **63,** 219.
Kaiser, D. G., Shaw, S. R., and Vangiessen, G. J. (1974). *J. Pharm. Sci.* **63,** 567.
Kelly, J. G., Nimmo, J., Rae, R., Shanks, R. G., and Prescott, L. F. (1973). *J. Pharm. Pharmacol.* **25,** 550.
Kelly, R. W. (1972). *In* "Proceedings of the International Symposium on Gas Chromatography and Mass Spectrometry" (A. Frigerio, ed.), p. 19. Tamburini, Milan.
Kowblansky, M., Scheinthal, B. M., Gravello, G. D., and Chafetz, L. (1973). *J. Chromatogr.* **76,** 467.
Kuksis, A. (1966). *In* "Methods of Biochemical Analysis" (D. Glick, ed.), p. 325. Interscience, New York.
Kupferberg, H. J. (1972). *J. Pharm. Sci.* **61,** 284.
Laik Ali, S. (1973). *Chromatographia* **6,** 478.
Lennard, M. S., Silas, J. H., Smith, A. J., and Tucker, G. T. (1977). *J. Chromatogr.* **133,** 161.
McCallum, N. K., and Armstrong, R. J. (1973). *J. Chromatogr.* **78,** 303.
Matin, S. B., and Rowland, M. (1972). *J. Pharm. Sci.* **61,** 1235.
Mee, J. M. L., and Brooks, C. C. (1971). *J. Chromatogr.* **62,** 138.
Mellström, B., and Ehrsson, H. (1974). *Acta Pharm. Suecica* **11,** 91.
Moffat, A. C., Horning, E. C., Matin, S. B., and Rowland, M. (1972). *J. Chromatogr.* **66,** 255.
Morvay, J. (1973). *Acta Med. Budapest* **29,** 145.
Nash, J. F., Bopp, R. J., and Rubin, A. (1971). *J. Pharm. Sci.* **60,** 1062.
Noonan, J. S., Blake, J. W., Murdick, P. W., and Ray, R. S. (1972). *Life Sci.* **11,** 363.
Omoto, S., Inouye, S., and Niida, T. (1971). *J. Antibiot.* **24,** 430.
Osman, M. A., and Hill, Jr., H. H. (1982). *J. Chromatogr.* **232,** 430.

Pellizzari, E. D., Brown, J. H., Talbot, P., Farmer, R. W., and Fabre, Jr., L. F. (1971). *J. Chromatogr.* **55,** 281.
Pero, R. W., Harvan, D., Owens, R. G., and Snow, J. P. (1972). *J. Chromatogr.* **65,** 501.
Pettit, B. C., and Stouffer, J. E. (1970). *J. Chromatogr. Sci.* **8,** 735.
Quilliam, M. A., and Westmore, J. B. (1978). *Anal. Chem.* **50,** 59.
Rajkowski, K. M., and Broadhead, G. D. (1972). *J. Chromatogr.* **69,** 373.
Richard, B. M., Manno, J. E., and Manno, B. R. (1974). *J. Chromatogr.* **89,** 80.
Roncucci, R., Simon, M-J., and Lambelin, G. (1971a). *J. Chromatogr.* **62,** 135.
Roncucci, R., Simon, M-J., and Lambelin, G. (1971b). *J. Chromatogr.* **57,** 410.
Rowe, E. L., and Machkovech, S. M. (1977). *J. Pharm. Sci.* **66,** 273.
Rühlmann, K., and Giesecke, W. (1961). *Angew. Chem.* **73,** 113.
Sabih, K., Klaassen, C. D., and Sabih, K. (1971). *J. Pharm. Sci.* **60,** 745.
Sakauchi, N., and Horning, E. C. (1971). *Anal. Lett.* **4,** 41.
Sakauchi, N., Kumaoka, S., and Hanawa, Y. (1972). *Endocrinology Jpn.* **19,** 589.
Schou, J., Steentoft, A., Worm, K., Moerkholdt, A. J., and Nielsen, E. (1971). *Acta Pharmacol. Toxicol.* **30,** 480.
Shamma, M., Deno, N. C., and Remar, J. F. (1966). *Tet. Lett.* **13,** 1375.
Simpson, P. M. (1973). *J. Chromatogr.* **77,** 161.
Smith, E. D., and Sheppard, Jr., H. (1965). *Nature* **208,** 878.
Smith, E. D., and Shewbart, K. L. (1969). *J. Chromatogr. Sci.* **7,** 704.
Solow, E. B., Green, J. B. (1971). *Clin. Chem. Acta* **33,** 87.
Stavinoha, W. B., and Ryan, L. C. (1965). *J. Pharmacol. Exptl. Therap.* **150,** 231.
Stenlake, J. B., Williams, W. D., and Skellern, G. G. (1970). *J. Chromatogr.* **53,** 285.
Stewart, J. T., Duke, G. B., and Willcox, J. E. (1969). *Anal. Lett.* **2,** 449.
Street, H. V. (1971). *Clin. Chim. Acta* **34,** 357.
Street, H. V. (1972). *J. Chromatogr.* **73,** 73.
Sweeley, C. C., Bentley, R., Makita, M., and Wells, W. W. (1963). *J. Am. Chem. Soc.* **85,** 2495.
Thomas, B. H., Solomonraj, G., and Coldwell, B. B. (1973). *J. Pharm. Pharmacol.* **25,** 201.
Umeh, E. O. (1970a). *J. Chromatogr.* **51,** 139.
Umeh, E. O. (1970b). *J. Chromatogr.* **51,** 147.
Umeh, E. O. (1971). *J. Chromatogr.* **56,** 29.
Van Giessen, B., and Tsuji, K. (1971). *J. Pharm. Sci.* **60,** 1068.
Vecchi, M., and Kaiser, K. (1967). *J. Chromatogr.* **26,** 22.
Venturella, V. S., Gualario, V. M., and Lang, R. E. (1973). *J. Pharm. Sci.* **62,** 662.
Verebely, K., and Inturrisi, C. E. (1973). *J. Chromatogr.* **75,** 195.
Walle, T. (1975). *J. Chromatogr.* **114,** 345.
Watson, E., and Kalman, S. M. (1971). *J. Chromatogr.* **56,** 209.
Watson, J. R., and Crescuolo, P. (1970). *J. Chromatogr.* **52,** 63.
Welton, B. (1970). *Chromatographia* **3,** 211.
Westley, J. W., and Halpern, B. (1966). *J. Org. Chem.* **33,** 2283.
Wickramasinghe, J. A. F., and Shaw, S. R (1972). *J. Chromatogr.* **71,** 265.
Wilcox, M. (1967). *J. Pharm. Sci.* **56,** 642.
Wilkinson, G. R., and Leong Way, E. (1969). *Biochem. Pharmacol.* **18,** 1435.
Wotiz, H. H., and Clark, S. J. (1966). "Gas Chromatography in the Analysis of Steroid Hormones." Plenum, New York.
Wotiz, H. H., Charrancsol, G., and Smith, I. N. (1967). *Steroids* **10,** 127.
Zomzely, C., Marco, G., and Emery, E. (1962). *Anal. Chem.* **34,** 1414.
Zumwalt, R. W., Kuo, K., and Gehrke, C. W. (1971). *J. Chromatogr.* **57,** 193.

Chapter 3

The Control of Purity and Stability of Pharmaceuticals

I. Introduction

A number of analytical techniques are employed to monitor the purity and stability of drugs: ultraviolet (UV), visible, infrared (IR), and nuclear magnetic resonance (NMR) spectrometry, and gas (GC or GLC), thin-layer (TLC), and high performance liquid chromatography (HPLC). The decision which technique or combination of techniques to use depends on the nature of the drug and the expected impurities or decomposition products. Gas chromatography, because of its sensitivity and high resolution, is well suited to monitor purity, stability, and moisture content and has even been used for particle-size and surface-area measurements. The use of large-diameter preparative columns in gas chromatography has introduced new standards of purity.

It is important to remember that a compound yielding a single peak on a gas chromatogram is not *necessarily* pure. It may be that there are polar or high molecular weight impurities present that are not readily chromatographed. For this reason a thorough knowledge of the impurities likely to be present is necessary before gas chromatography can be employed, and thin-layer chromatography and ultraviolet spectroscopy are often carried out in parallel.

II. Control of Purity

The versatility of gas chromatography in quality control is best illustrated by specific examples.

A. Contaminants of Phenacetin

Although phenacetin has now been withdrawn in a number of countries, it provides a good illustration of the technique. During the 1950s there were reports of toxicity associated with phenacetin, and suggestions were made that impurities introduced during the synthesis could be responsible. It was much later before good evidence appeared. (Schnitzer, 1971). The three impurities most likely to be present were acetanilide, *p*-chloroacetanilide, and *p*-phenetidine. The U.S. Pharmacopeia described qualitative tests for *p*-phenetidine and acetanilide, and set a maximum limit of 0.03% for *p*-chloroacetanilide using paper chromatography. In order to improve on this, Koshy and co-workers (1965) produced a gas chromatographic method that could measure the three impurities in the presence of phenacetin.

Phenacetin itself was assayed by dissolving 250 mg in 25 ml of chloroform containing 250 mg of internal standard, benzocaine. An injection of 5 μl of this solution was made onto the column. The column was an aluminum tube 2 ft long and $\frac{1}{4}$ in in diameter and contained 4% Epon 1001 on silanized Chromosorb G (70–80 mesh). Injection-port and detector temperatures were 260 and 240°C, respectively. To measure phenacetin, acetanilide, and *p*-chloroacetanilide, the column was operated at a temperature of 188°C with helium as carrier gas at a rate of 100 ml/min. To detect *p*-phenetidine, the column was run at 145°C for 3 min, then the temperature was increased at a rate of 15°C/min to 180°C, and held for 20 min. Flame ionization detection was used.

Under these conditions the retention times were as follows:

Compound	t_R (min)
p-Phenetidine	3
Acetanilide	7
Benzocaine	9.5
p-Chloroacetanilide	12
Phenacetin	17.5

The authors went on to examine eight commercial samples of phenacetin of unknown age obtained from various sources. All samples contained 0–10 ppm of *p*-phenetidine and acetanilide; Only two contained less than 0.03% of *p*-chloroacetanilide, the upper limit set by the USP XVII. By using GLC, a simple and sensitive method for measuring three different impurities simultaneously was possible.

B. Purity of Phenylbutazone in Raw Material

Phenylbutazone has been reported as being unstable under normal storage conditions in tablets, suppositories, and solutions for injection (Beckstead *et al.*, 1968). Watson and colleagues (1973), using GLC, found that phenylbutazone chromatographed as a single peak on 5% OV-7 at 150°C. Although the decomposition products produced on storage broke down in the injection port of the chromatograph, a reproducible pattern of artifact peaks could be obtained. The method involved mixing a weighed quantity of raw material, equivalent to about 100 mg of phenylbutazone, with 2 ml water, 0.25 ml concentrated hydrochloric acid, and 10 ml ethyl acetate. The mixture was centrifuged, the organic phase was diluted with a solution containing 4 mg of diphenyl phthalate as internal standard, and 2 μl of the sample solution was injected. The column size was 6 ft × 6 mm; it was packed with 5% OV-7 on Gas Chrom Q (100–120 mesh) and was operated isothermally at a temperature of 150°C with nitrogen (60 ml/min) as the carrier gas. The injection-port and detector temperatures were 230 and 240°C, respectively. Flame ionization was used. The retention times obtained were

Compound	t_R (min)
Artifact Peaks	4, 4.8, 6.0, 7.9, and 24.2
Phenylbutazone	32
Diphenyl phthalate	44

Six different dosage forms were examined and the authors found that a peak with retention time of 4 min was common to all. A possible end product of phenylbutazone degradation is azobenzene, and this is eluted, under the above conditions, as two peaks with retention times of 4 and 7.9 min.

The extraction of phenylbutazone from official preparations did not present a problem, and the authors found that acetone or sodium hydroxide could be used. Difficulty was anticipated with preparations containing buffering agents since the C-4 hydrogen of phenylbutazone is labile and, in the presence of aluminum hydroxide or magnesium carbonate, the drug might be complexed as a salt and its quantitative extraction prevented. For this reason the acid medium described above was used.

C. Trace Impurities in Halothane

The anesthetic gas halothane, 2-bromo-2-chloro-1,1,1-trifluoroethane, may contain traces of impurities the nature and biological properties of which have aroused interest. The most difficult problem is the choice of stationary phases that will permit a separation of the impurities from halothane and also allow each to be resolved from the others in order that quantitative measurements can be made.

Chapman *et al.* (1967) selected four phases: (i) chlorinated biphenyl, (ii) dinonyl phthalate, (iii) poly(ethylene glycol) 400, and (iv) a column of dinonyl phthalate followed by a larger column of poly(ethylene glycol). The chlorinated biphenyl column (30% Aroclor 1254 on Chromosorb P operated at 60°C) allowed the identification of nine compounds eluted before halothane and four after. Analysis using the other columns revealed a total of 16 compounds that could be possible contaminants of halothane. The chlorinated biphenyl column and the dinonyl phthalate linked to poly(ethylene glycol) columns were sufficient to allow identification and quantitation of 14 impurities. However, the remaining two, 2-bromo-1,1,1-trifluoroethane and 2,2-dichloro-1,1,1-trifluoroethane, were eluted together. Normally it is not necessary to separate the two, but it can be done using poly(ethylene glycol) alone on a Celite support. The authors concluded that by using their method no impurity other than those identified by themselves would be present at a concentration greater than 10 ppm.

D. Solvents in Steroid Hormones

Many steroids prepared synthetically contain traces of organic solvent trapped in the crystal lattice, and evidence of this can be found by microscopic examination or as loss of weight after drying. A detailed study of this problem was made, using 5% poly(ethylene glycol) 400 on Chromosorb W acid washed (60–80 mesh) at 30°C (Rasmussen *et al.*, 1972a). The authors used a thermostated water bath, rather than a conventional oven, to maintain a temperature of 30°C. In a later paper by the same authors (Rasmussen *et al.*, 1972b), a solid injector at 230°C was used, and hydrolysis of some specimens was noted due to the relatively long injection time of approximately 30 sec. For example, when cortisol acetate was injected by this technique a compound was eluted from the column having the same retention time as acetic acid. For quantitative, rather than qualitative, determination of the trapped solvents, a Porapak Q column (2 m × 2 mm) at 175°C was used. Methanol, ethanol, acetone, methyl ethyl ketone, and ethyl acetate were detected in a number of steroids as shown in Table 3.1.

Table 3.1

Organic Solvents Identified in Samples of Synthetic Steroids[a]

Steroid	Solvent	μg Solvent/mg steroid
Deoxycorticosterone acetate	Ethyl Acetate	0.16
Estradiol	Ethanol	0.18
Estradiol benzoate	Ethanol	0.62
Hydrocortisone	Methyl ethyl ketone	1.05
Hydrocortisone acetate	Acetone	0.12
Methyl testosterone	Ethyl acetate	0.66
Prednisolone	Acetone	1.84
Prednisone	Methanol	0.62
	Acetone	2.02
Progesterone	Methanol	0.15

[a] From Rasmussen *et al.* (1972b).

E. Purity of Hexadiphane

Hexadiphane, 1,1-diphenyl-3-hexamethyleneiminopropane (**I**) is used as an antispasmodic agent and is obtained by decyanation of 1,1-diphenyl-3-hexamethyleneiminobutyronitrile (**II**)

$(H_5C_6)_2CH\,CH_2CH_2N$ (I) ← $(H_5C_6)_2C(C{\equiv}N)\,CH_2CH_2N$ (II)

In order to make sure that quantitative conversion has been achieved, a rapid and specific method of measuring both compounds is needed. A successful method has been developed (Mardente and De Marchi, 1975) using a stainless steel column (2 m × 2 mm) containing 0.5% OV-17 on Gas Chrom Q (80–100 mesh). The column was operated at 190°C for 6 min, then heated to 250°C at a rate of 22°C/min, and maintained at that temperature. The injection-port and detector temperatures were 300 and 280°C, respectively, and nitrogen at 35 ml/min was used as the carrier gas.

Hexadiphane hydrochloride (25 mg) was dissolved in 100 ml of water to give a standard solution, and diphenylpyroline hydrochloride (16 mg) in 100 ml of water served as an internal standard solution. To measure the purity of solid material, it was first finely powdered, and a quantity approximately equivalent to 4 mg of hexadiphane hydrochloride or 6 mg hexadiphane maleate was transferred to a 50-ml volumetric flask, and 40 ml of

0.1 *M* HCl was added. The flask was heated to between 70 and 80°C and shaken for 30 min. The solution was then cooled, made up to volume with 0.1 *M* HCl, mixed, and centrifuged. A 5-ml sample of this solution was mixed with 3 ml of the internal standard solution, made basic with one or two drops of 1 *M* NaOH, and extracted with 10 ml of diethyl ether followed by a further extraction with 5 ml. After centrifugation the ether was removed, and the residue was dried over sodium sulfate and evaporated to dryness. The residue was dissolved in 0.2 ml of carbon disulfide, and a 2-μl portion was injected. The retention times were as follows: diphenylpyroline, 3 min; hexadiphane, 5 min; and 1,1-diphenyl-3-hexamethyleneiminobutyronitrile, 9 min.

The authors used the method to assay a liquid preparation and a combination with oxazepam hemisuccinate and compared the results with those obtained by titration with sodium laurylsulfate. The results obtained were very similar, although the standard deviation of the GC method was slightly greater. The authors concluded that the GC method was simple, rapid, and selective, and they point out that the titrimetric method would not reveal the presence of the precursor without an additional separation step.

F. Determination of *N,N*-Dimethylaniline in Penicillins and Cephalosporins

One route to the semisynthetic penicillins uses 6-aminopenicillanic acid as an intermediate, and, at one stage of the synthesis, an organic base is added to remove hydrogen chloride, which is produced. *N,N*-Dimethylaniline is often used as this base, and it is necessary to monitor the final product for residual traces, particularly in view of the role of aromatic amines in cancer. In the method developed by Margosis (1977) the bulk penicillin (1 g) was made alkaline with 5 ml of 5% NaOH, and the amine was extracted with 1 ml of cyclohexane containing naphthalene as internal standard. A portion of the cyclohexane phase was injected directly onto a 6-ft × 2-mm column of 3% OV-17 on Gas Chrom Q at 60°C. The injection and detector temperatures were both ~150°C. Helium at 30 ml/min was used as the carrier gas, and detection was by flame ionization. Under these conditions *N,N*-dimethylaniline had a retention time of 8 min, and that of the internal standard was 16 min. The method was precise and sensitive: at the 1.6-ppm level, a coefficient of variation of less than 8% was easily obtained.

The method was applied to 6-aminopenicillanic acid from four manufacturers and to a number of penicillins and cephalosporins. The 6-aminopenicillanic acid samples all contained significant amounts of *N,N*-dimethylaniline, ranging from 60 to almost 800 ppm. Except for a few instances with ampicillin, the final penicillins and cephalosporins were relatively free of the amine. The author states that the situation with the ampicillin has been largely rectified due to improved purification or use of a less potentially harmful base. No *N,N*-dimethylaniline (i.e., <0.15 ppm) was detected in samples of cephradine, cephaloglycin, cefazolin, penicillamine, penicillin G procaine, penicillin G potassium, penicillin, oxacillin, cloxacillin, dicloxacillin, and carbenicillin.

G. Detection of Contamination Arising from Sterilization of Pharmaceutical Products with Ethylene Oxide

Ethylene oxide is employed extensively to gas-sterilize manufactured goods such as plastics, food, and pharmaceutical products, and a number of studies have been published, using GLC, to demonstrate the range of contaminants arising from its use. Residual ethylene oxide was first demonstrated to be mixed with steroids, using gas chromatography with thermal-conductivity detection (Adler, 1965). Later work showed that ethylene oxide residues were present in sterilized plastics (O'Leary *et al.*, 1969), and irradiated poly(vinyl chloride) (PVC), subsequently sterilized with ethylene oxide, was found to contain 2-chloroethanol (Whitbourne *et al.*, 1969). In this latter study, 2-chloroethanol was separated by vacuum extraction, which was claimed to be superior to water extraction which, in its turn, was better than acetone extraction.

Methyl methacrylate powder is a component of bone cement kits, and the presence of residual ethylene oxide after sterilization has been reported (Zugar, 1972). In that method, the methyl methacrylate powder was swirled in a flask with acetone, and, after 10 min, 1 ml of the vapor phase was injected onto a 3-ft stainless steel column packed with a styrene–divinylbenzene copolymer (Chromosorb 101) maintained at 120°C. FID detection was used, and helium at 30 ml/min was the carrier gas. Under these conditions, ethylene oxide had a retention time of 2 min and emerged ahead of the acetone, which had a retention time of 3.5 min. Using this method, ethylene oxide was detected at the 1 ppm level. The author recommended that high grade or redistilled acetone be used to avoid a peak that is eluted before the ethylene oxide.

Another and more rapid method using aqueous injection and flame ionization has been used to study ethylene oxide and its reaction products, ethylene chlorohydrin and ethylene glycol, in bulk drugs, aqueous suspensions, and ointments (Hartman and Bowman, 1977). Gas chromatography was carried out using 180-cm × 3-mm glass columns, the packings being

Porapak P for ethylene oxide and 15% Carbowax 400 on Gas Chrom Q (80–100 mesh) for ethylene chlorohydrin and ethylene glycol. Column temperature was ~110°C for all, and helium at 60 ml/min was used as the carrier gas. Aqueous solutions were chromatographed directly, aqueous suspensions were centrifuged, and a portion of the supernatant was injected. Water-insoluble bulk drugs were shaken with distilled water for 5 min, centrifuged, and a portion of the clear supernatant was injected directly. Ointments (3 g) were shaken with water-washed hexane (6 ml). Exactly 3.0 ml of distilled water was added, and mixing was carried out for 5 min, followed by centrifugation. The organic upper layer was aspirated, and the aqueous layer was used for injection. Recoveries were greater than 90% and precision was high. The authors claim that water provided the highest extraction efficiency and eliminated any interference in the separation that would have arisen if an organic solvent had been used.

III. Drug Stability

A. Aspirin Decomposition during Storage

Gas chromatography can be used to check the stability of drugs themselves by demonstrating the presence of decomposition products. For example, it is important to monitor the salicylic acid content of aspirin formulations since aspirin can decompose during storage (Watson *et al.*, 1972). The USP XVIII raised the limit for salicylic acid in compressed aspirin tablets from 0.15% to 0.30% and from 0.75% to 3.0% in coated and buffered tablets. It has also become apparent that aspirin is less stable in aspirin–propoxyphene formulations (p. 57) than in the more common formulations, and the National Formulary XIII set an upper limit of 3% for free salicylic acid in capsules of aspirin–propoxyphene. Watson and colleagues derivatized their aspirin formulations with diazomethane, which converts acetylsalicylic acid to its methyl ester and free salicylic acid to methyl salicylate. Separation was achieved on a 6-ft glass column of 6% OV-17 on Gas Chrom Q (100–120 mesh). The column was run at 135°C for 8 min, and then the temperature raised by 2°C/min until the internal standard methyl *o*-methoxybenzoate was eluted. The retention times of methyl salicylate and methyl-*o*-methoxybenzoate were 7 and 16 min, respectively. The time of contact between aspirin and the diazomethane–tetrahydrofuran was kept to a minimum to reduce the possibility of aspirin hydrolysis. The method was found to be precise and more rapid than the Trap Column method with spectrophotometric determination (Levine and Weber, 1968).

A gas chromatographic study of tablets and granulations containing aspirin has also been carried out (Patel *et al.*, 1972). The material was extracted with anhydrous diethyl ether, and the residue on evaporation was silylated with BSA. The resulting mixture was chromatographed on a 3-ft × 4-mm glass column packed with 1% OV-210. The injection-port and detector temperatures were 175 and 350°C, respectively, and the column was held at 100°C for 8.5 min before being heated to 230°C at a rate of 15°C/min. Under these conditions the TMS ether of salicylic acid had a retention time of 6 min. There was also an "unknown" peak with a much longer retention time (18 min). This was suspected to be acetylsalicylsalicylic acid, and chromatography of the synthesized material on two columns of different polarity confirmed this. It was demonstrated that acetylsalicylsalicylic acid was not formed on column, and it was concluded that it must have arisen during the manufacturing process.

The percentage of salicylic acid found in commercial aspirin tablets and granulates ranged from traces (less than 0.02%) to 0.17%. Acetylsalicylsalicylic acid content ranged from 0.02 to 0.1%. The percentage of salicylic acid found in propoxyphene–aspirin-type analgesic combinations ranged from 0.2 to 30.0% and in codeine–aspirin analgesics from 0.1 to 34.5%. Laik Ali (1974) has attempted the measurement of salicyclic acid in the presence of acetylsalicylic acid using methyl iodide–potassium carbonate in acetone to form the methyl esters. He found that to measure salicylic acid in bulk aspirin, a column-chromatographic separation of salicylic acid was needed prior to derivatization. In a more recent study (1975) he compared BSTFA, MSTFA, and methyl iodide as derivatizing agents. The silyl derivatives were found to be more sensitive, but, again, slight hydrolysis of aspirin to salicylic acid during derivatization was observed, and a preliminary column separation was necessary.

B. Decomposition of Amitriptyline Hydrochloride in Aqueous Solution

The antidepressant amitriptyline hydrochloride can be formulated as a solution and sterilized by filtration according to the BP 1974. The stability of amitriptyline during autoclaving and on long term storage has been studied by GLC (Enever *et al.*, 1975). Preliminary work showed that decomposition could take place when solutions of amitriptyline in water or phosphate buffer were autoclaved at 115–116°C in the presence of excess oxygen. The contents of an autoclaved ampule were extracted with diethyl ether (1 ml) and 5 μl was injected onto a 5-ft × 0.6-cm glass column packed with 3% OV-25 on Chromosorb W AW/DMCS (80–100 mesh) operated at 220°C. Nitrogen at 28.5 ml/min was used as carrier gas, and flame ionization detection was employed. TLC and UV spectroscopy of the degradation products were also carried out, and identification was made by GC–MS and NMR.

In addition to the peak for amitriptyline, two other peaks were obtained on GLC, and three spots on TLC.

Compound	t_R (min)
Decomposition product	
A	3.5
B	4.5
Amitriptyline	7.0

Mass spectrometry and nuclear magnetic resonance studies identified **A** and **B** as 3-(propa-1,3-dienyl)-1,2:4,5-dibenzocyclohepta-1,4-diene and dibenzosuberone. The third decomposition product (**C**), detected only on TLC, was found to be 3-(2-oxoethylidine)-1,2:4,5-dibenzocyclohepta-1,4-diene. The authors proposed the following pathway of decomposition:

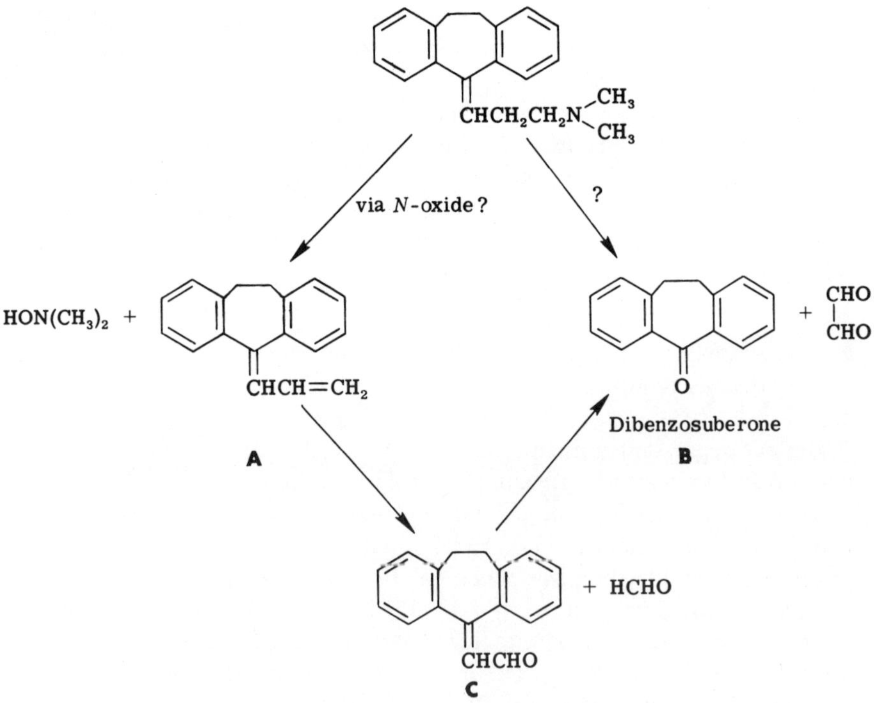

$CHCH_2CH_2N(CH_3)_2$

via *N*-oxide?

?

$HON(CH_3)_2$ + $CHCH{=}CH_2$

A

O + CHO–CHO

Dibenzosuberone

B

+ HCHO

CHCHO

C

The formation of dibenzosuberone was postulated to occur during autoclaving by direct oxidation of the olefinic double bond. Compound **A** was postulated as arising via the *N*-oxide and a loss of dimethylhydroxylamine; further oxidation would yield **C**.

IV. Determination of Water Content

The accurate measurement of the water content of ingredients making up pharmaceutical preparations and of the final products themselves is very important. The water content of many substances used in pharmacy, such as starches, vegetable drugs, oils, granulates, tablets, and ointments can be determined either directly or indirectly by gas chromatography using a thermal conductivity detector. A review of the early work done in this area has been published by Johnson (1967). The tailing of the water peak in much of the earlier work has been greatly reduced by the advent of polymeric fluorocarbon packings such as Haloport F, Fluoropak, and the Porapak vinylbenzene polymers.

A detailed Swiss study has compared three different techniques for estimating the moisture content of ointments, granulates, and tablets (Iconomou *et al.*, 1969). The three techniques were (i) loss of weight on drying, (ii) titration with Karl Fischer reagent, and (iii) gas chromatography. The gas chromatographic method needed 0.1 mg of substance, which was transferred to a 50-ml standard flask; 1-propanol was added as internal standard, and then the solution was diluted to the mark with water-free methanol. After standing for 4–5 h, 5 μl was injected onto a 3-m × 2.5-mm glass column containing a 10% coating of a heptaglycol–monoisononylphenyl ether on teflon powder (35–60 mesh). The column was operated at 65°C, and the injection-port and detector temperatures were 180 and 240°C, re-respectively. The retention times of water and 1-propanol were 8 and 12 min, respectively. The results are summarized in Table 3.2. The authors concluded that gas chromatography is useful, especially in cases where the "loss on drying" technique cannot be used, e.g., with thermolabile substances, and with substances that react with the Karl Fischer reagent.

Another study (Rasmussen *et al.*, 1971) compared the use of gas chromatography and Karl Fischer titration to determine the moisture content of tablets of potassium phenoxymethylpenicillin. It was found that the Karl Fischer method apparently included the theoretical quantity of water of crystalization in the lactose excipient, while GC did not. This demonstrates that GC can differentiate between water present as moisture and water present in a chemically bound form. The authors suggest that the Karl Fischer method be used for total water, and that gas chromatography or Karl Fischer reagent applied to a methanol extract be used for adsorbed water.

Table 3.2

Determination of Water Content[a]

Substance	Water content (%)		
	Loss on drying	Karl Fischer	GC
Unguentum refrigerans, Ph.Helv.V. (spermaceti cream)	17.99	—	18.20
Lanolinum Ph.Helv.V. (lanolin)	20.21	—	20.28
Granulatum simplex KA (simple granulate)	6.33	9.42	9.88
Granulatum phenacetini KA (phenacetin granulate)	0.83	7.10	8.76
Compressi acidi acetylsalicylici compositi KA (compound acetyl salicylic acid tablets)	0.88	3.10	3.27
Compressi allobarbitali compositi KA (compound allobarbital tablets)	2.31	3.33	3.61
Compressi analgetici compositi KA (compound analgesic tablets)	2.49	3.80	3.89
Ascorbinsaüre brausetabletten (effervescent ascorbic acid tablets)	1.10	—	1.27

[a] From Iconomou *et al.* (1969).

V. Measurement of Surface Areas of Powders by Gas Chromatography

Gas solid chromatography can be used to study and measure the surface area of powders (see Fig. 3.1). In this technique the powder to be studied is *itself* the column packing, and the carrier gas, often a mixture of nitrogen and helium, is passed in the normal way. After equilibrium has been reached

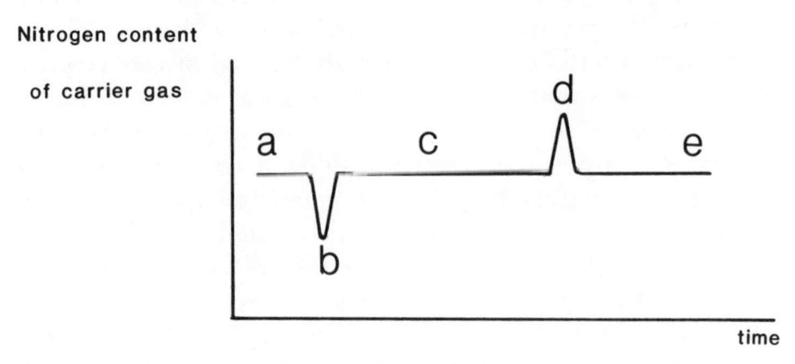

FIG. 3.1. Determination of surface area of a powder by packing it in a column and measuring the change in composition of a carrier gas mixture (see text).

(a), the column is rapidly cooled using a Dewar flask filled with liquid nitrogen. The nitrogen in the carrier gas is adsorbed onto the material filling the column, and the change in composition of the carrier gas can be measured by a thermal conductivity detector (b). When adsorption is complete, the composition of the carrier gas returns to its original proportions (c). If the Dewar flask is then removed and the column heated rapidly to its initial temperature, the nitrogen will be desorbed, giving a change in composition of carrier gas (d). Once all the nitrogen has been released, the baseline will revert to its original position (e).

The amount of nitrogen adsorbed can be calculated from the height of the peaks d and b, and the surface area of the powder obtained from the isotherm equation developed by Brunauer *et al.* (1938). This is a highly specialized use of gas chromatography. Volumetric or gravimetric methods have also been used to measure the amount of gas adsorbed, and indeed, some of the instruments and techniques used are difficult to classify as gas chromatography. However, the potential for using GC in this field exists, even though it does not seem to be used frequently. The surface areas of tranexamic acid, calcium carbonate, cloxacillin sodium, levodopa, and microcrystalline cellulose have been studied (Nystrom *et al.,* 1977), and an instrument for pharmaceutical powders has been described (Johansson and Stanley-Wood, 1977).

References

Adler, N. (1965). *J. Pharm. Sci.* **54,** 735.

Beckstead, H. D., Kaistha, K. K., and Smith, S. J. (1968). *J. Pharm. Sci.* **57,** 1952.

Brunauer, S., Emmett, P. H., and Teller, E. (1938). *J. Am. Chem. Soc.* **60,** 309.

Chapman, J., Hill, R., Muir, J., Suckling, C. W., and Viney, D. J. (1967). *J. Pharm. Pharmacol.* **19,** 231.

Enever, R. P., Li Wan Po, A., Millard, B. J., and Shotton, E. (1975). *J. Pharm. Sci.* **64,** 1497.

Hartman, P. A., and Bowman, P. B. (1977). *J. Pharm. Sci.* **66,** 789.

Iconomou, N., Seth, P. L., and Büchi, J. (1969). *Pharm. Acta Helv.* **44,** 433.

Johansson, M. E., and Stanley-Wood, N. G. (1977). *Powder Technol.* **16,** 145.

Johnson, C. A. (1967). *In* "Advances in Pharmaceutical Sciences," Vol. 2, p. 224. Academic Press, London and New York.

Koshy, K. T., Wickersham, H. C., and Duvall, R. N. (1965). *J. Pharm. Sci.* **54,** 1547.

Laik Ali, S. (1974). *Chromatographia* **7,** 655.

Laik Ali, S. (1975). *Chromatographia* **8,** 33.

Levine, J., and Weber, J. D. (1968). *J. Pharm. Sci.* **57,** 631.

Mardente, S., and De Marchi, F. (1975). *J. Pharm. Sci.* **64,** 1866.

Margosis, M. (1977). *J. Pharm. Sci.* **66,** 1634.

Nystrom, C., Malmqvist, K., and Wulf, A. (1977). *Acta Pharm. Suecica* **14,** 497.

O'Leary, R. K., Watkins, W. D., and Guess, W. L. (1969). *J. Pharm. Sci.* **58,** 1007.

Patel, S., Perrin, J. H., and Windheuser, J. J. (1972). *J. Pharm. Sci.* **61,** 1794.

Rasmussen, K. E., Waaler, T., Karlsen, J., and Baerheim Svendsen, A. (1971). *Pharm. Acta Helv.* **46,** 53.

Rasmussen, K. E., Rasmussen, S., and Baerheim Svendsen, A. (1972a). *J. Chromatogr.* **69,** 370.

Rasmussen, K. E., Rasmussen, S., and Baerheim Svendsen, A. (1972b). *J. Chromatogr.* **71,** 542.

Schnitzer, B. (1971). *Ann. Intern. Med.* **75,** 320.

Watson, J. R., Matsui, F., Lawrence, R. C., and McConnell, P. M. J. (1973). *J. Chromatogr.* **76,** 141.

Watson, J. R., Matsui, F., McConnell, P. M. J., and Lawrence, R. C. (1972). *J. Pharm. Sci.* **61,** 929.

Whitbourne, J. E., Mogenhan, J. A., and Ernst, R. R. (1969). *J. Pharm. Sci.* **58,** 1024.

Zugar, L. A. (1972). *J. Pharm. Sci.* **61,** 1801.

Chapter 4

Analysis of Excipients, Preservatives, and Related Compounds in Pharmaceutical Preparations

We have seen how gas chromatography can be used to monitor impurities in pharmaceutical preparations, but it can, of course, also be used to measure excipients, preservatives, and the therapeutically active substances themselves. This chapter examines the use of gas chromatography for the first two classes of compounds, and the next chapter considers the determination of active substances.

I. The Analysis of Waxes

Many natural and synthetic waxes are used in ointments and cosmetic preparations, and GLC is often used to characterize the various fractions after a preliminary separation by column or thin-layer chromatography. Abate *et al.* (1970) examined beeswax and candelilla wax and found that the hydrocarbons of beeswax consist mainly of odd carbon chains (C_{23}–C_{33}) with a predominance of C_{27}, while the hydrocarbons of candelilla consisted of C_{29}–C_{33} with C_{33} predominant. The alcohols and straight-chain fatty acids of the waxes were also studied, the acids being chromatographed as methyl esters after derivatization with diazomethane. The authors proposed GLC as a means of detecting adulteration of a wax.

The wax from the alfalfa plant *Stipa tenacissima* L. (Gramineae) was analyzed by a group of French workers, using a combination of gas chromatography and mass spectrometry (GC–MS) (Miet *et al.*, 1972). Approximately 70% consisted of hydrocarbons, 10% free fatty acids, and the remainder esters and aliphatic alcohols. The hydrocarbons were isolated using column chromatography and the free acids by ion exchange. Gas chromatography was carried out on a 95-cm × 3.2-mm column packed with 10% SE-52. The free acids were methylated using diazomethane and separated on the same column.

Another group of French workers reported an unusual opportunity to combine gas chromatography with history (Jung *et al.*, 1972). In 1969, the demolition of some old buildings in the Rue des Grandes-Arcades in Strasbourg revealed a trench in which were found some wooden vessels. These were dated as thirteenth or fourteenth century and contained a pale yellow odorless substance. Preliminary chemical tests revealed that the substance was almost entirely organic in nature, less than 3% being inorganic. An elemental analysis was made, yielding the following: carbon 72.74%, hydrogen 9.97%, nitrogen 1.13%. Oxygen analysis contributed a further 14.98%. Fatty acids were extracted, methylated with diazomethane, and chromatographed on a 2-m column of 10% DEGS on Chromosorb G, operated at 100°C, and then programmed to 190°C at 5°C/min. The chromatogram obtained is shown in Fig. 4.1.

The presence of erucic acid indicated a vegetable oil of the Cruciferae family. Further examination by thin-layer chromatography and NMR revealed the presence of cholesterol and wool fat. The authors concluded that the presence of wool fat was responsible for the preservation of the ointment for 6 or 7 centuries.

Barry and Saunders (1971) studied cetostearyl alcohol, a base for a number of cosmetic creams and lotions. They used a stainless-steel column (2 m

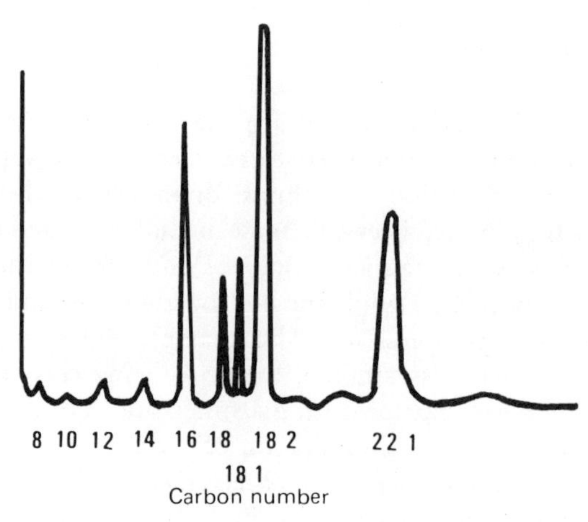

FIG. 4.1. Chromatogram of methylated fatty acids derived from residues found in some thirteenth- or fourteenth-century wooden vessels. From Jung *et al.* (1972). Reproduced with permission of *Ann. Pharm. Franc.*

× 0.125 in) coated with 10% KOH on acid-washed Chromosorb W (60–80 mesh) and operated at 240°C. The composition was found to be octadecanol 72.8%, hexadecanol 24.7%, dodecanol 1.3%, and tetradecanol 0.5%. There were also traces of decanol and two unidentified impurities.

Poly(ethylene glycols) (PEG) are also used in ointments and cosmetic milks, and these can also be readily measured by gas chromatography (Laurent *et al.*, 1973). A 6-ft × 5-mm column packed with 3% SE-20 on Chromosorb AW DMCS was used with temperature programming from 200 to 350°C at 15°C/min. A number of different poly(ethylene glycols) could be identified using thin-layer chromatography, but to distinguish between PEG 300 and 400, it was necessary to gas chromatograph the methylated derivatives. Both PEG 300 and 400 gave rise to a large number of peaks, but only those with retention times greater than 11 min were used for identification. This technique enabled PEG 400 to be identified in ointments based on niflumic acid and hydrocortisone acetate.

In a detailed series of studies of oils and waxes, Fawaz *et al.* (1973) described the analysis of lanolin. Preliminary TLC and column chromatography was carried out before GC–MS. Lanolin was found to consist of sterols and triterpene alcohols (68.3%), 17.1% monohydroxy alcohols, and 8.7% alkanediols. A total of 38 monohydroxy alcohols and 31 alkanediols were identified.

II. Ethanol in Drug Preparations

Ethanol is present in many pharmaceutical products such as tinctures, essences, elixirs, linaments, and toilet preparations. Until the advent of gas chromatography the standard procedure of quantitation was to distil the ethanol from the liquid being tested and measure the specific gravity of the distillate. Often a purification step was needed before a meaningful specific gravity could be obtained. During the 1960s a great deal of progress was made in the gas chromatographic determination, which was more rapid, selective, and sensitive. This early work is reviewed in two detailed studies of the gas chromatographic determination of ethanol in drug preparations (Ellis and Wragg, 1970; Harris, 1970).

Ellis and Wragg compare the distillation procedure with a gas chromatographic method using propanol as an internal standard. The column used was 4 m × 2.2 mm, stainless steel filled with 5% Nonidet P40 on Teflon 6 (30–40 mesh). The column was operated at 70°C with nitrogen as the carrier gas. Examples of chromatograms, using flame ionization and thermal conductivity, were included. The thermal conductivity detector, of course, also measures water. The precision of the gas chromatographic method was determined by applying it to the analysis of pharmaceutical preparations ranging from Figs Syrup [compound British Pharmaceutical Codex (BPC), 4.1% ethanol v/v] to Iodine Solution (simple BPC 1959, 93.3% ethanol v/v). The standard deviation ranged from 0.03 to 0.18. The precision of the official methods was also measured and found to be similar to that of gas chromatography, although the official methods tended to give slightly lower results. This tendency was observed to be most pronounced with samples containing higher concentrations of ethanol. The authors went on to present results of ethanol determinations by both GC and official methods for 101 drug preparations. These are presented in Table 4.1.

The study concluded with a comparison of GC and official methods for the determination of alcohol in methylated drug preparations. The official methods gave a measure of the total alcohol present and made no distinction between methanol and ethanol. The gas chromatographic method generally yielded results closely comparable to those obtained by the official methods, except for some preparations where the use of one of the more complex official methods produced less accurate results. The precision of both techniques was similar.

These general conclusions are supported by the work of Harris (1970) who used a 6-ft × 1/8-in column packed with Porapak Q (80–100 mesh) and operated at 160°C. Flame ionization was the method of detection. Under these conditions methanol, ethanol, isopropyl alcohol, and 1-propanol could be readily separated with retention times of 1.5, 2.5, 3.5, and 5 min, respectively. Distillation and gas chromatography were compared by measuring the spirit content over the range 1.8 to almost 160% proof spirit. The results are presented in Table 4.2. It can be seen that the chromatographic method gives, in many cases, a slightly higher result.

It would be a pity to end this section on alcohol measurement without mentioning the method published to measure alcohol in "alcohol-free" beverages (Morad *et al.*, 1980). Many alcohol-free beers are produced from water, malt, and hops, the fermentation being prematurely arrested. Alternatively, the alcohol content of "normal" beer is reduced by distillation or reverse osmosis.

The authors used amyl alcohol as the internal standard and added it to the beverage to be tested. The flask was inverted several times, allowed to stand for 10 min, and a portion was injected onto a 10% PEG 20M column, 1.5 m × 4 mm, on 100–120 mesh diatomite MAW. Flame ionization was used, and the column temperature was 150°C. Nitrogen at a flow rate of 30 ml/min was used as the carrier gas. Under these conditions the retention times of ethanol and amyl alcohol were 42 and 90 sec, respectively. On applying this method to a range of alcohol-free beers, grape, and apple juices, the authors found that all contained ethanol in concentrations ranging from

Table 4.1

Results of the Determination of Ethanol in Drug Preparations[a]

No.	Preparation	Ethanol (%, v/v)				
		Official method	Incorporated	By GLC	By official method	Difference (GLC minus official method)
1	Cetrimide Solution, Strong, B.P.C.	I	7.1	7.1	7.3	−0.2
2	Nux Vomica Elixir, B.P.C.	I	—	10.3	10.1	0.2
3	Saccharin Solution, B.P.C. 1954	I	11.3	11.4	11.5	−0.1
4	Hamamelis Water, B.P.C.	I	—	13.3	13.4	−0.1
5	Liquorice Liquid Extract, B.P.	I	—	17.9	17.3	0.6
6	Cascara Liquid Extract, B.P.	I	—	21.4	21.2	0.2
7	Quassia Infusion, Concentrated, B.P.C. 1959	I	—	21.7	21.7	0
8	Buchu Infusion, Concentrated, B.P.C. 1954	I	—	21.9	21.2	0.7
9	Morphine Hydrochloride Solution, B.P.	Id	22.6	22.6	23.0	−0.4
10	Valerian Infusion, Concentrated, B.P.C. 1963	I	—	22.9	22.1	0.8
11	Strychnine Hydrochloride Solution, B.P.C. 1963	Id	22.6	23.0	23.0	0
12	Ipecacuanha Tincture, B.P.	Id	—	25.4	25.1	0.3
13	Quillaia Liquid Extract, B.P.C.	I	—	32.2	31.6	0.6
14	Stramonium Liquid Extract, B.P.	I	—	36.7	36.2	0.5
15	Nux Vomica Liquid Extract, B.P.	I	—	38.5	38.7	−0.2
16	Senega Liquid Extract, B.P.C.	Ig	—	38.9	39.1	−0.2
17	Catechu Tincture, B.P.C.	I	—	39.8	39.4	0.4
18	Gentian Tincture, Compound, B.P.	I	—	42.4	42.3	0.1
19	Stramonium Tincture, B.P.	I	—	43.9	43.4	0.5
20	Nux Vomica Tincture, B.P.	I	43.5	44.0	43.5	0.5
21	Quillaia Tincture, B.P.C.	I	43.9	44.0	43.7	0.3
22	Brilliant Green and Crystal Violet Paint, B.P.C.	I	45.3	45.2	44.6	0.6
23	Opium Tincture, B.P.	I	—	45.6	45.9	−0.3

(*Continued*)

Table 4.1 ***(Continued)***

No.	Preparation	Ethanol (%, v/v)				
		Official method	Incor- porated	By GLC	By official method	Difference (GLC minus official method)
24	Rhubarb Tincture, Compound, B.P.	I	—	50.6	50.9	−0.3
25	Belladonna Liquid Extract, B.P.C.	I	—	50.7	50.2	0.5
26	Quinine Solution, Ammoniated, B.P.C. 1963	If	53.2	52.9	53.3	−0.4
27	Senega Tincture, B.P.C.	Ig	55.9	56.3	56.3	0
28	Hyoscyamus Liquid Extract, B.P.	I	—	56.9	56.3	0.6
29	Gelsemium Tincture, B.P.C.	I	—	58.7	58.2	0.5
30	Lobelia Tincture, Simple, B.P.C. 1949	I	—	58.8	58.1	0.7
31	Colchicum Tincture, B.P.	I	67.8	67.1	67.1	0
32	Hyoscyamus Tincture, B.P.	I	68.8	69.1	68.6	0.5
33	Belladonna Tincture, B.P.	I	—	68.1	68.2	−0.1
34	Iodine Solution, Strong, B.P. 1958	Ic	78.0	78.0	77.4	0.6
35	Iodine Solution, Weak, B.P.	Ic	87.6	88.0	87.2	0.8
36	Iodine Solution, Simple, B.P.C. 1959	Ic	93.1	93.3	92.7	0.6
37	Diamorphine and Pine Elixir, Compound, B.P.C. 1949	II	22.6	22.9	22.0	0.9
38	Rose Water, Concentrated, B.P.C. 1949	II	44.9	45.1	44.8	0.3
39	Cinnamon Water, Concentrated, B.P.C.	IIa	54.0	53.8	52.9	0.9
40	Chloroform Water Concentrated, B.P.C. 1959	II	54.0	54.0	53.5	0.5
41	Caraway Water, Concentrated, B.P.C.	II	54.4	54.1	53.5	0.6
42	Peppermint Water, Concentrated, B.P.	II	53.9	54.2	53.2	1.0
43	Camphor Water, Concentrated, B.P.C. 1949	II	53.9	54.3	53.2	1.1

Table 4.1 (*Continued*)

		Ethanol (%, v/v)				
No.	Preparation	Official method	Incor-porated	By GLC	By official method	Difference (GLC minus official method)
44	Orange Flower Water, Concentrated, B.P.C. 1949	II	53.9	54.3	53.4	0.9
45	Dill Water, Concentrated, B.P.C.	II	53.9	54.4	53.2	1.2
46	Camphor Liniment, Ammoniated, B.P.C.	IV	55.7	55.7	55.6	0.1
47	Opium Tincture, Camphorated, Concentrated, B.P.C. 1963	II	—	56.2	54.1	2.1
48	Opium Tincture, Camphorated, B.P.	II	58.8	59.1	57.6	1.5
49	Ether Spirit, B.P.C.	IIa	63.0	63.0	61.3	1.7
50	Anise Water, Concentrated, B.P.C.	II	63.1	63.4	62.8	0.6
51	Benzaldehyde Spirit, B.P.C.	II	72.4	72.4	71.6	0.8
52	Peppermint Spirit, B.P.C.	II	80.0	80.3	78.3	2.0
53	Camphor Spirit, B.P.C. 1959	II	80.5	80.7	79.0	1.7
54	Chloroform Spirit, B.P.	IIb	85.7	85.8	85.6	0.2
55	Lemon Spirit, B.P.C.	II	85.7	86.1	83.7	2.4
56	Nitrous Ether Spirit, B.P.C. 1959	II	—	86.5	85.0	1.5
57	Orange Spirit, Compound, B.P.C.	II	88.0	88.0	86.6	1.4
58	Ammonia Solution, Aromatic, B.P.C.	IIIf	3.4	3.4	3.5	−0.1
59	Figs Syrup, Compound, B.P.C.	III	—	4.1	4.3	−0.2
60	Senna Syrup, B.P.C. 1959	III	5.4	4.9	4.8	0.1
61	Senna Mixture, Compound, B.P.C. 1959	IIIf	8.5	8.3	8.3	0
62	Paracetamol Elixir, Paediatric, B.P.C.	III	11.2	11.1	11.3	−0.2
63	Senna Elixir, B.P.C. 1949	III	12.4	12.1	12.4	−0.3
64	Ephedrine Elixir, B.P.C.	III	12.5	12.4	12.4	0
65	Chloroform and Morphine Tincture, B.P.C.	III	13.5	14.1	13.5	0.6

(*Continued*)

Table 4.1 (*Continued*)

		Ethanol (%, v/v)				
No.	Preparation	Official method	Incor-porated	By GLC	By official method	Difference (GLC minus official method)
66	Diamorphine and Terpin Elixir, B.P.C.	III	18.1	18.3	18.6	−0.3
67	Squill Linctus, Opiate, B.P.C.	IIId	18.8	18.7	18.9	−0.2
68	Senna Liquid Extract, B.P.C.	III	—	20.0	19.4	0.6
69	Rhubarb Infusion, Concentrated, B.P.C. 1959	III	—	20.5	19.7	0.8
70	Senna Infusion, Concentrated, B.P.C. 1959	III	—	21.8	21.2	0.6
71	Gentian Infusion, Compound, Concentrated, B.P.	III	—	21.8	20.8	1.0
72	Flexible Collodion, B.P.	IIIa	21.7	22.4	20.4	2.0
73	Seneg Infusion, Concentrated, B.P.C.	IIIf	—	22.7	22.4	0.3
74	Tolu Solution, B.P.C.	III	—	23.3	22.4	0.9
75	Orange Peel Infusion, Concentrated, B.P.C.	III	—	23.6	23.1	0.5
76	Soap Spirit, B.P.C.	IIIf	32.8	33.0	32.1	0.9
77	Hamamelis Liquid Extract, B.P.C.	III	—	34.0	34.1	−0.1
78	Squill Liquid Extract, B.P.C.	III	—	36.6	36.1	0.5
79	Euphoroia Liquid Extract, B.P.C. 1949	III	—	37.9	37.3	0.6
80	Phenobarbitone Elixir, B.P.C.	III	38.3	38.2	37.7	0.5
81	Arnica Flower Tincture, B.P.C. 1949	III	—	45.3	44.4	0.9
82	Valerian Liquid Extract, B.P.C. 1963	III	—	45.6	44.4	1.2
83	Valerian Tincture, Ammoniated, B.P.C. 1963	IIIf	—	53.1	52.3	0.8
84	Cardamom Tincture, Compound, B.P.	III	—	54.7	53.3	1.4
85	Squill Tincture, B.P.C.	III	—	56.2	54.9	1.3

Table 4.1 (*Continued*)

No.	Preparation	Ethanol (%, v/v) Official method	Incor-porated	By GLC	By official method	Difference (GLC minus official method)
86	Valerian Tincture, Simple, B.P.C. 1949	III	—	57.0	55.7	1.3
87	Cocillana Liquid Extract, B.P.C.	III	—	57.1	55.6	1.5
88	Capsicum Tincture, B.P.C.	III	—	60.1	58.7	1.4
89	Lobelia Tincture, Ethereal, B.P.C.	IIIa	—	61.7	59.8	1.9
90	Digitalis Tincture, B.P.C. 1959	III	—	63.7	62.6	1.1
91	Orange Tincture, B.P.C.	III	—	66.7	65.4	1.3
92	Ipecacuanha Liquid Extract, B.P.	III	—	68.7	67.2	1.5
93	Ammonia Spirit, Aromatic, B.P.C.	IIIf	—	69.9	69.3	0.6
94	Benzoin Tincture, Compound, B.P.C.	III	—	74.2	73.1	1.1
95	Tolu Tincture, B.P.C. 1959	III	82.9	82.9	81.1	1.8
96	Benzoin TIncture, B.P.C.	III	—	83.8	82.8	1.0
97	Ginger Tincture, Strong, B.P.	III	—	86.7	84.9	1.8
98	Myrrh Tincture, B.P.C.	III	—	86.8	85.3	1.5
99	Ginger Tincture, Weak, B.P.	III	89.9	90.4	89.1	1.3
100	Cardamom Tincture, Aromatic, B.P.C.	III	—	91.0	89.1	1.9
101	Salicylic Acid Lotion, B.P.C.	III	92.8	92.9	91.8	1.1
			Mean of differences			0.75

[a] From Ellis and Wragg (1970). Reproduced with permission of The Royal Society of Chemistry.

0.009 to 0.385% v/v (beer), 0.022 to 0.146% v/v (grape juice), and 0.019 to 0.03% v/v (apple juice). The authors suggest that the large difference in ethanol content of the beers may be due to the inefficiency of the technique of alcohol removal. It is also possible that ethanol could be produced during storage if microorganisms were not completely destroyed.

Table 4.2

Comparison of Results Obtained by Distillation and Gas Chromatography[a]

Sample	Proof spirit (%)		Difference, $D - G$
	Distillation (D)	Gas chromatography (G)	
1	1.85	1.75	0.1
2	7.65	7.65	0
3	17.55	17.85	−0.3
4	45.75	45.45	0.3
5	46.8	46.8	0
6	56.9	56.0	0.9
7	58.5	59.6	−1.1
8	72.55	72.3	0.25
9	80.5	81.15	−0.65
10	102.2	100.7	1.5
11	117.35	118.4	−1.05
12	136.1	134.8	1.3
13	152.05	153.7	−1.65

[a] From Harris (1970). Reproduced with permission of The Royal Society of Chemistry.

III. Sorbitol and Mannitol

Sorbitol and mannitol are used as bulking agents in some preparations. Sorbitol has been measured in irrigating solution by converting it to the more volatile hexaacetate derivative using acetic anhydride–pyridine. Chromatography was carried out on a 3% QF-1 column operated at 220°C (Helgren *et al.*, 1972). Sorbitol hexaacetate had a retention time of 7 min, and the internal standard, bis(2-ethylhexyl) sebacate, 13 min. The authors recommended that their GC method be evaluated as a replacement for the USP adsorption column chromatography procedure.

In a different study, Manius *et al.* (1972) determined sorbitol and mannitol together in aqueous solution as the hexaacetates with dioctyl sebacate as internal standard. A stainless-steel column, 0.61 m × 0.32 cm, packed with 5% ECNSS-M on Gas Chrom Q (80–100 mesh), was operated at 175°C. The TMS esters were also made, but no resolution could be obtained on columns of OV-17, QF-1, XE-60, Dow 710, and PEG 4000 MS. The authors found their method simpler to perform, more specific, and less time consuming than the USP method. The retention times obtained were as follows: dioctyl sebacate 8 min, mannitol hexaacetate 13 min, and sorbitol hexaacetate 19 min. Although the authors used dioctyl sebacate as internal standard they suggested that dinonyl sebacate might be better since it would be eluted closer to the mannitol derivative. They were, however, unable to obtain pure dinonyl sebacate. Six replicate assays of a 40% sorbitol solution had a mean concentration of 45.1% with a standard deviation of ±1.15%.

IV. Antioxidants and Preservatives

Antioxidants are incorporated in a large number of pharmaceutical preparations in order to prolong the useful life of the active ingredients and to prevent deterioration of susceptible compounds. For example, if no antioxidants were added to preparations containing oils or fats, these would become unacceptable rapidly because of the development of rancidity. Preservatives are added to prevent the growth of bacteria in sterile preparations and ophthalmic solutions.

An early study of the gas chromatography of a number of antioxidants (Choy *et al.*, 1963) examined several stationary phases. The most satisfactory was found to be a column of 20% SE-30 on Chromosorb W (60–80 mesh) at 185°C. The antioxidants BHA (2- and 3-*tert*-butyl-4-hydroxyanisole), BHT (3,5-di-*tert*-butyl-4-hydroxytoluene), and ethoxyquin (1,2-dihydro-6-ethoxy-2,2,4-trimethylquinoline) were studied, and the results of the analysis of BHA and BHT in vitamin A oil and a multiple vitamin mixture were given.

Gosselé (1971) studied the gas chromatography of TMS derivatives of sorbic and benzoic acids as well as the methyl, ethyl, and propyl esters of *p*-hydroxybenzoic acid on 3% SE-30 with temperature programming from 90–290°C. In the same year, a detailed Japanese study (Takemura, 1971) examined the chromatographic properties of 15 synthetic preservatives and 7 antioxidants as TMS ethers on several stationary phases with *n*-hexadecane as internal standard and using thermal conductivity detection (see Tables 4.3 and 4.4). Takemura found that nonpolar phases such as Apiezon L and SE-31 were preferable to the more polar phases for the separation of the antioxidants and preservatives as their TMS ethers. A great deal of tailing was observed with the polar columns, and some compounds could not be detected. Recoveries appeared to be high, 99.9 and 100.4% being reported for MHB and PHB, respectively. A typical chromatogram is illustrated in Fig. 4.2, and it can be seen that a good separation is achieved. All but one of the compounds eluted within 10 min.

Table 4.3

Stationary Phases and Operating Conditions[a]

Phase	Support	Length (m)	Temp (°C)	Carrier gas flow rate (ml/min)
20% Apiezon L		2.63	200	30.1
20% SE 31		2.63	200	22.0
20% DC 550	Celite	2.63	200	24.7
20% Ucon oil 50 HB	545	1.87	180	22.1
280 X	(60–80			
15% Poly(diethylene glycol) succinate	mesh)	2.63	160	15.9
20% Carbowax 20 M		2.63	160	23.6

[a] From Takemura (1971). Adapted with permission of The Japanese Society for Analytical Chemistry.

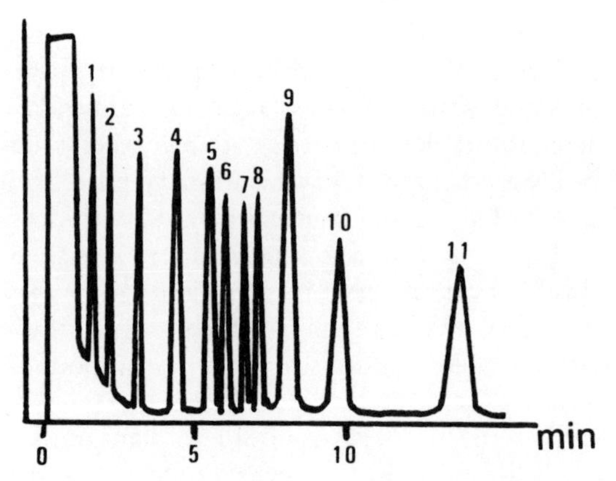

FIG. 4.2. Chromatogram on 20% Silicone SE-31. 1, SA; 2, BA; 3, CMP; 4, CX; 5, MHB; 6, BHT; 7, BHA; 8, EHB; 9, internal standard; 10, PHB; 11, BHB. From Takemura (1971). Reproduced with permission of The Japanese Society for Analytical Chemistry.

Table 4.4

Retention Times of Antioxidants and Preservatives Relative to *n*-Hexadecane[a]

		Apiezon L	Carbowax 20 M	PDEGS	Ucon oil 50HB-280X	Silicone DC-550	Silicone SE-31
Preservatives	Methyl-4-hydroxybenzoate (MHB)	0.538	—	2.225	1.265	0.861	0.697
	Ethyl-4-hydroxybenzoate (EHB)	0.702	—	2.439	1.754	1.258	0.876
	n-Propyl-4-hydroxybenzoate (PHB)	1.043	—	3.143	2.487	1.790	1.224
	n-Butyl-4-hydroxybenzoate (BHB)	1.595	—	4.386	3.867	2.614	1.729
	4-Chloro 3-methylphenol (CMP)	0.309	—	0.495	0.509	0.465	0.403
	4-Chloro-3,5-xylenol (CX)	0.446	1.610	0.665	0.724	0.659	0.545
	2-Hydroxydiphenyl	0.940	5.466	2.248	1.857	1.563	1.049
	Phenylsalicylate	3.032	—	—	—	5.772	2.901
	2,5-Di-*tert*-butylhydroquinone	1.052	1.941	0.750	1.767	2.009	1.817
	Sorbic acid (SA)	0.119	—	—	—	0.242	0.192
	Salicylic acid	0.468	—	—	—	0.955	0.738
	Benzoic acid (BA)	0.204	—	0.625	—	0.368	0.286
	1,3-Dihydroxy 1-hexylbenzene	1.603	4.203	1.441	3.029	2.714	2.207
	5-Methyl-2-isopropyl-1-phenol	0.236	0.662	—	0.376	0.399	0.363
	1,1,1-Trichloro-2-methyl-2-propanol	0.148	—	—	—	0.232	0.234
Antioxidants	3-*tert*-Butyl-4-hydroxyanisole (BHA)	0.623	2.806	1.145	1.236	1.085	0.826
	2,6-Di-*tert*-butyl 4-methylphenol (BHT)	0.658	3.288	1.378	1.165	—	0.751
	Methyl-3,4,5-trihydroxybenzoate	1.236	—	2.673	3.854	3.251	2.355
	Ethyl-3,4,5-trihydroxybenzoate	1.814	—	3.606	5.333	4.636	3.204
	n-Propyl-3,4,5-trihydroxybenzoate	2.661	—	4.838	7.564	6.553	4.393
	i-Amyl-3,4,5-trihydroxybenzoate	—	—	—	—	—	7.148
	Ethyl-3,4-dihydroxybenzoate	1.276	—	3.105	—	2.667	1.842
n-Hexadecane		15.9 min	8.9 min	4.4 min	7.8 min	8.9 min	8.0 min

[a] From Takemura (1971). Reproduced with permission of The Japanese Society for Analytical Chemistry.

References

Abate, V., Badoux, V., Hicks, S. Z., and Messinger, M. (1970). *J. Soc. Cosmetic Chem.* **21,** 119.

Barry, B. W., and Saunders, G. M. (1971). *J. Pharm. Sci.* **60,** 645.

Choy, T. K., Qualtrone, J., and Alicino, N. J. (1963). *J. Chromatogr.* **12,** 171.

Ellis, J. R., and Wragg, J. S. (1970). *Analyst* **95,** 16.

Fawaz, F., Chaigneau, M., and Puisieux, F. (1973). *Ann. Pharm. Franc.* **31,** 217.

Gosselé, J. A. W. (1971). *J. Chromatogr.* **63,** 429.

Harris, J. R. (1970). *Analyst* **95,** 158.

Helgren, P. F., Thomas, M. A., and Theivagt, J. G. (1972). *J. Pharm. Sci.* **61,** 103.

Jung, L., Mélais, M-Cl., and Bachoffner, P. (1972). *Ann. Pharm. Franc.* **30,** 205.

Laurent, F., Hommel, D., and Jung, L. (1973). *Ann. Pharm. Franc.* **31,** 601.

Manius, G., Mahn, F. P., Venturella, V. S., and Senkowski, B. Z. (1972). *J. Pharm. Sci.* **61,** 1831.

Miet, C., Fawaz, F., Choix, M., and Puisieux, F. (1972). *Ann. Pharm. Franc.* **30,** 263.

Morad, A. M., Hikal, A. H., and Buchanin, R. (1980). *Chromatographia* **13,** 161.

Takemura, I. (1971). *Bunseki Kagaku* **20,** 62.

Chapter 5

The Determination of Therapeutically Active Substances in Pharmaceutical Preparations

This chapter reviews the gas chromatographic methods that have been used, and are being used, to measure drug substances in various types of pharmaceutical preparations. Of course, a wide range of other analytical techniques are available to do the same job, e.g., infrared, ultraviolet and visible spectrometry, nuclear magnetic resonance spectrometry, polarography, and thin-layer and high performance liquid chromatography. Gas chromatography is probably most useful when low concentrations of non-fluorescent or non-UV-absorbing substances must be measured. When the drug substance constitutes a considerable fraction of the preparation, then it may be simpler, and cheaper, to use one of the other methods listed above. Methods for the gas chromatographic analysis of drug substances in their pharmaceutical preparations are outlined below, and the therapeutically active substances have been grouped under their main pharmacological action for convenience.

I. Anti-Inflammatory and Analgesic Agents

Several methods have been published for the analysis of aspirin in dosage forms, and mention of some of them has already been made in Chapter 3. Aspirin is easily extracted and is chromatographed well when converted to the methyl ester. Use of 10% methyltrifluoropropyl silicone (Patel *et al.*, 1972) or 3% OV-17 (Laik Ali, 1974) has been reported.

More recently the analysis of the propionic acid-derived anti-inflammatory, fenclorac, has been described (Visalli *et al.*, 1976). A suitable quantity of tablets or the contents of a number of capsules were ground to a fine powder and the equivalent of 125 mg of fenclorac (**I**) was weighed into a 50-ml volumetric flask and 30 ml of methylene chloride added. After shaking for 5 min the solution in the flask was made up to the mark with more methylene chloride. A portion of this solution was centrifuged, and a 2-ml aliquot of the supernatant was mixed with 1 ml of internal standard, triphenylethylene (125 mg in 50 ml of methylene chloride), and 0.2 ml of *N,O′*-bis(trimethylsilyl)acetamide in a vial. The vial was shaken and allowed to stand at room temperature for at least 15 min before injecting a 2-μl portion onto the column of the gas chromatograph. The glass column, 4 ft × 2 mm, was packed with 5% OV-25 on Chromosorb W HP (80–100 mesh). The column was operated isothermally at 195°C, nitrogen was used as the carrier gas at 30 ml/min, and flame ionization was the method of detection. The column was conditioned with several injections of derivatized standard before the test solution was injected. Under these conditions the retention times of the TMS esters of fenclorac and the internal standard were 5.5 and 8 min, respectively.

Recovery was good, ranging from 94.6 to 104.9%, and the detector response to fenclorac was linear over the range 0.31–3.13 mg/ml. Precision was also very good: the mean of six determinations had a coefficient of variation of 0.65%. The method could also be used to measure 3-chloro-4-cyclohexylphenylglycolic acid (**II**) and the diethylamine salt of α-chloro-4-cyclohexylphenylacetic acid (**III**), impurities associated with the synthesis of fenclorac. Compound **II** is also a degradation product of fenclorac. Both compounds could be detected down to concentrations of 0.01% in **I**.

An earlier paper by Dechene *et al.* (1969) described the analysis of acetylsalicylic acid and phenacetin in the presence of caffeine and codeine in APC and codeine tablets. A suitable number of tablets were powdered and weighed to yield about 450 mg of acetylsalicylic acid and 325 mg of phenacetin. The powder was placed in a 25-ml volumetric flask containing 20 ml of chloroform. The flask was shaken for about 15 min at 50°C, then cooled and made up to the mark with more chloroform. The contents of the flask were filtered, the first 5 ml being discarded, and a 2-μl sample of the clear filtrate was injected onto the column. A stainless steel 6-ft × $\frac{1}{8}$-in column packed with 2% Dow Corning No. 200 on Haloport F-80 was used and operated initially at 100°C, then rising to 180°C at 10°/min. Nitrogen at 50 ml/min was used as the carrier gas, and flame ionization was used for detection. Under these conditions, the retention times of acetylsalicylic acid and phenacetin were 4.1 and 9.6 min, respectively. For 10 assays of APC tablets, a mean recovery of 98.3% for aspirin and 100.77% for phenacetin was obtained with standard deviations of 2.77 and 3.26% respectively.

II. Antihistamines

The gas chromatography of amine mixtures in drug formulations has been studied by Hishita and Lauback (1969). Phenylpropanolamine, phenylephrine, phenyltoloxamine, and chlorpheniramine were determined in tablets, capsules, and syrups after conversion to their TMS derivatives. Tablets or capsules containing about 40 mg phenylpropanolamine hydrochloride, 10 mg phenylephrine hydrochloride, 15 mg phenyltoloxamine dihydrogen citrate, and 5 mg chlorpheniramine maleate were shaken with 50 ml distilled water in a bath maintained at 100°C. After cooling and centrifugation, a 15-ml aliquot was shaken with 4.5 g of sodium carbonate, than 4 g each of sodium chloride and sodium sulfate were added, and the flask was shaken again. Extraction was then performed with three 15-ml portions of isopropyl alcohol and the combined extracts were diluted to 100 ml with more isopropyl alcohol. The alcoholic phase was then centrifuged, and a 5-ml sample was transferred to a serum vial. Alcoholic HCl (0.16 ml) was added, the solution was evaporated to dryness in a stream of nitrogen, and the residue was redissolved in 0.4 ml of internal standard (tribenzylamine in pyridine, 2 mg/ml). Finally, 0.1 ml of BSA was added, the mixture was stirred at room temperature, and a 2-μl sample was injected onto a 2-m × 4-mm glass column packed with 0.1% DC 710 on 60–80 mesh DMCS-treated glass beads. The column was programed from 100 to 200°C at 10°C/min. Helium at 80 ml/min was the carrier gas, and the injection port and detector temperatures were 300 and 260°C, respectively.

For syrups, a 10-ml sample was diluted to 15 ml with water and shaken with 4.5 g of sodium carbonate. The resulting solution was then treated as for tablets except that 10 μl of trifluoroacetic acid was added in place of 0.16 ml of alcoholic hydrochloric acid. The use of alcoholic hydrochloric acid to form salts was satisfactory for tablets and capsules but was found to lead to incomplete silylation and extraneous peaks when used with syrups. Trifluoroacetic acid could also be used with tablets and capsules and, indeed, was found to be more satisfactory since too much hydrochloric acid resulted in the degradation of phenylephrine, while too little led to the degradation of phenylpropanolamine. No degradation was observed when trifluoroacetic acid was used. Under the chromatographic conditions outlined above, the retention times of the TMS derivatives obtained were as follows: phenylpropanolamine 3 min, phenylephrine 5.6 min, phenyltoloxamine 7.9 min, chlorpheniramine 8.3 min, and tribenzylamine (internal standard) 10.8 min. The method was found to be precise, and there was close agreement between the label claim and the "found" concentration except with phenyltoloxamine, which was found to be approximately 8% lower than the stated concentration.

Rader (1969) produced a method for measuring pyrilamine and methapyriline in tablets also containing chlorpheniramine. He chose to measure the chlorpheniramine by UV spectrophotometry and the other two by gas chromatography. A column packed with *p*-toluenesulfonic acid was used to extract the amines from the tablets. Pyrilamine and methapyriline were eluted from the column with chloroform, then 1% acetic acid in chloroform was used to elute the chlorpheniramine. The eluate containing pyrilamine and methapyriline was reduced in volume to about 1 ml then diluted in a volumetric flask to 5 ml with methanol, and 5–6 μl was injected onto a 6-ft × 4 mm-glass column packed with 10% SE-52 on Hi EFF-8 BP (80–100 mesh) operated at 250°C. Thermal conductivity detection was used, and the carrier gas was nitrogen at a flow rate of 50 ml/min. No retention times were given, but the results from six commercial samples and two synthetic mixtures were presented. The recoveries for pyrilamine and methapyriline ranged from 95 to 100% and 100 to 105%, respectively. Among the other compounds present in some of the commercial samples were caffeine, phenylpropanolamine hydrochloride, salicylamide, ephedrine sulfate, and atrophine sulfate.

Chlorpheniramine and phenylpropanolamine in cough–cold preparations have been measured without derivatization (Mario and Meehan, 1970). A low-load nonpolar column of 2% SE-30 was chosen because of the range of polarity of the ingredients present in some of the preparations, e.g., glyceryl guaiacolate and dextromethorphan). A 25-ml portion of cough syrup was made acid by adding 5 ml concentrated hydrochloric acid then shaken with carbon tetrachloride (3 × 30 ml). The organic phase was washed with 30 ml deionized water, and then the carbon tetrachloride was discarded. The deionized water wash was combined with the original aqueous phase, and 25 ml 50% sodium hydroxide was added. The aqueous phase, now alkaline, was extracted with chloroform (7 × 30 ml). This seems rather excessive, but no comment was made on the reason for this by the authors. In addition, 5 g sodium chloride was added to the combined organic phases during the first four extractions. The total chloroform extract was filtered and reduced in volume (never to dryness) on a steam bath. A 5-ml volume of internal standard solution (pramoxine hydrochloride—1 g extracted from aqueous solution and diluted to 100 ml with chloroform) was added, and the resulting solution was diluted to 50 ml with more chloroform. A 2-μl sample was injected onto a glass column (8 ft × 0.125 in) packed with 2% SE-30 on Chromosorb W HP (80–100 mesh). The column was operated isothermally at 180°C, and helium at a flow rate of 30 ml/min was used as the carrier gas. Under these conditions the retention times were as shown in Table 5.1.

Table 5.1

Retention Times of Constituents of Cough–Cold Preparations[a]

Compound	t_R (min)
Phenylpropanolamine	1
Glyceryl guaiacolate	3.5
Chlorpheniramine	10
Dextromethorphan	15
Pramoxine	25

[a] From Mario and Meehan (1970).

All peaks showed tailing. The analysis of cough mixtures presents problems because of the polarities of some of the compounds present. Selection of an internal standard with a shorter retention time would reduce the chromatography time. The phenylpropanolamine tailing could be reduced by derivatization, but chlorpheniramine, a tertiary amine, could not be derivatized simply.

Wong *et al.* (1973) reported a simple and rapid method of measuring meclozine hydrochloride in tablets. A quantity of tablets equivalent to about 100 mg of meclozine hydrochloride was pulverized, weighed, and transferred to a 125-ml flask. A 50-ml volume of chloroform containing 50 mg of dinonyl phthalate was added, and the mixture was shaken for 30 min. After centrifugation, 5 μl of the chloroform solution was injected onto a 6-ft × 0.25-in stainless steel column packed with 3% OV-17 on Chromosorb W AW DMCS (80–100 mesh) operated at 290°C. Helium at 75 ml/min was used as the carrier gas with flame ionization as the detection system. Under these conditions the internal standard, dinonyl phthalate, and meclozine had retention times of 7.5 and 14.3 min, respectively. The average recovery was 100.4% with a precision of 1.7%.

Finally, a paper has appeared that describes the measurement of salicylamide, phenylpropanolamine hydrochloride, caffeine, chlorpheniramine maleate, phenylephrine hydrochloride, and pyrilamine maleate simultaneously in capsule preparations (De Fabrizio, 1980). The contents of a number of capsules were mixed and a sample corresponding to about 240 mg salicylamide was heated at 80°C for 15 min with 30 ml ethanol in a 50-ml volumetric flask. The flask was then shaken until cool, 5 ml internal standard added (80 mg dicyclohexyl phthalate in 50 ml chloroform), and the contents of the flask were made up to the mark with ethanol. An aliquot was centrifuged, and 0.5 ml was evaporated to dryness under nitrogen at room

Table 5.2

Resolution of Constituents of Capsule Preparations[a]

	mg/capsule	
	Mean found	Theoretical
Salicylamide	241	240
Phenylpropanolamine	9.30	9.25
Caffeine	40.6	40
Chlorpheniramine maleate	1.56	1.5
Phenylephrine HCl	5.82	5.75
Pyrilamine maleate	7.91	7.8

[a] From De Fabrizio (1980).

temperature. The residue was treated with 0.1 ml of a mixture of 1.2% 4-dimethylaminopyridine in pyridine and acetic anhydride (1 : 1) for 18 min at 40°C. One μl of the reaction mixture was then injected onto a glass column (1.8 m × 2 mm) packed with 8% OV-101 on Chromosorb W HP (80–100

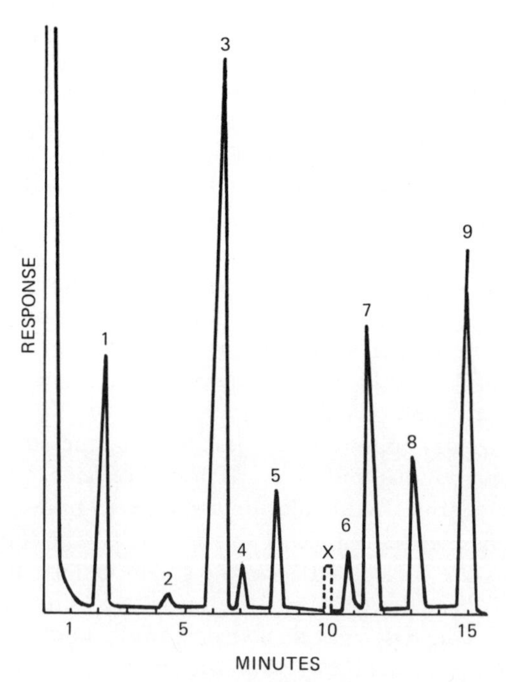

FIG. 5.1. Representative gas chromatogram. Key: 1 and 2, reagent peaks; 3, salicylamide; 4, phenpropanolamine; 5, caffeine; 6, chlorpheniramine; 7, phenylephrine; 8, pyrilamine; and 9, internal standard. From De Fabrizio (1980), reproduced with permission of the copyright owner.

mesh) programed from 170 to 270°C at 9°C/min after an initial 3 min at the lower temperature. Nitrogen at 15 ml/min was the carrier gas, and flame ionization detection was used. Under the derivatization conditions described above, acetyl derivatives of salicylamide, phenylpropanolamine, and phenylephrine were formed. A good resolution of all the constituents was obtained (Fig. 5.1). Recoveries were high (97.7–99.7%), and the mean "found" concentrations corresponded closely to the theoretical (Table 5.2). The procedure was simple and accurate, the six active ingredients being chromatographed in 17 min.

III. Hypnotics, Sedatives, and Tranquilizers

Neckopulos (1971) measured barbiturates in tablets by simple solvent extraction, followed by direct injection of the free acids. After pulverizing 20 tablets, about 150 mg of the powder was accurately weighed into a 50-ml beaker. The powder was then transferred to a separating funnel containing dilute hydrochloric acid and shaken with chloroform (4 × 10 ml). The chloroform was filtered each time through anhydrous sodium sulfate into a 50-ml round-bottom flask and evaporated to dryness. A 10-ml volume of an internal standard solution (150 mg hexabarbital in 100 ml of methanol) was added, and a 5-μl sample was injected onto a stainless steel column packed with 3% OV-17 on Gas Chrom Q (60–80 mesh). The column was run for 3 min at 150°C and then programed to 210°C at a rate of 4°C/min. Helium at 60 ml/min was used as the carrier gas. No retention times were given, but the order of elution was buto-, pento-, seco-, hexo-, and, finally, phenobarbital. A single lot of Nidar® tablets (containing the sodium salts of buto-, pento-, seco-, and phenobarbital) was analyzed twenty times (Table 5.3).

The percentage recoveries for the various barbiturates were as follows: buto- (100.8), pento- (100.2), seco- (100.3), and pheno- (98.4). Good agreement was found between this method and the Kjeldahl method for total barbiturate; however, the GC method was five times as rapid.

A GLC procedure for the assay of sedative and anticonvulsant compounds in capsules, tablets, oral suspensions, and injections has been reported (Watson *et al.*, 1978). For ethotoin, glutethimide, mephenytoin, methsuximide, and phensuximide, ten capsules were emptied, and the contents were weighed and thoroughly mixed. A sample of the powder, equivalent to about 8–12 mg of drug, was weighed into a 10-ml flask and shaken with 5 ml ethyl acetate, and 1 ml internal standard (diphenyl phthalate, 15 mg/ml in dimethylformamide). A quantity was injected directly onto the column of the gas chromatograph. For oral suspensions, a quantity equiva-

Table 5.3

Analysis of Nidar® Tablets Containing Barbiturates[a]

	Buto-	Pento-	Seco-	Pheno-
Mean	7.68	24.95	24.74	7.67
SD	0.16	0.44	0.59	0.17

[a] From Neckopulos (1971). Reproduced from the *Journal of Chromatographic Science* by permission of Preston Pubs. Inc.

lent to 600 mg of drug was diluted to 50 ml in a volumetric flask, and 1 ml of this was extracted with three 5-ml portions of ethyl acetate. These were combined, reduced in volume to 5 ml, and then 1 ml of diphenyl phthalate internal standard was added. A portion of this solution was injected directly.

In order to estimate primidone and phenytoin, derivatization was necessary. In the case of primidone, a sample equivalent to about 10–15 mg of drug was weighed into a 10-ml flask, 5 ml of dimethylformamide was then added, followed by 1.5 ml of *N,O*-bis(trimethylsilyl)acetamide. One ml of internal standard solution was added, and the mixture was shaken for 30 min at 60°C, and then a sample of a few microliters was injected. A quantity of phenytoin or its sodium salt, equivalent to about 10–15 mg, was weighed into a separatory funnel, and 3 ml of 0.1 *M* NaOH was added. When dissolution was complete, the solution was made acid with concentrated hydrochloric acid and extracted with ethyl acetate. The organic phase was evaporated, the residue was dissolved in 5 ml of dimethylformamide, and 1.5 ml of *N,O*-bis(trimethylsilyl)acetamide was added. A 1-ml portion of internal standard was added, and the mixture was shaken briefly and allowed to stand at room temperature for 15 min before injection.

A glass column (6 ft × 4 mm) was used for gas chromatography and was packed with 5% OV-101 on Chromosorb 750 (100–120 mesh) operated at 150°C for 5 min and then programed to 240°C at 3°C/min. The retention times were as given in Table 5.4. Results of the analysis of 19 dosage forms are presented, and, in general, a close agreement was found between the label claim and that found by GLC.

The specific determination of meprobamate in bulk and tablets has been reported (Rabinowitz *et al.*, 1972). Tablets or powder were ground, and a quantity, equivalent to ~40 mg of meprobamate, was accurately weighed. The powder was then shaken with 20 ml of methanol, centrifuged, evaporated to dryness, and dissolved in 1 ml of tybamate internal standard (2 mg/ml in methylene chloride). Approximately 1 μl was injected onto a glass column (76 cm × 3 mm) packed with 3.8% OV-17 on Gas Chrom Q (80–100 mesh). Nitrogen at 45 ml/min was used as the carrier gas, and the

Table 5.4

Retention Times of Anticonvulsants[a]

Drug	Retention time (min)
Methsuximide	5.86
Phensuximide	6.21
Ethotoin	10.86
Mephenytoin	11.20
Glutethimide	11.69
Primidone	11.75
Phenytoin	23.56
Diphenyl phthalate	29.20

[a] From Watson *et al.* (1978). Reproduced with permission of the copyright owner.

column was operated at 170°C with flame ionization detection. To condition the column, the following procedure was adopted:

1. 1 h at 250°C with carrier gas flowing, then cool to room temperature;
2. stop carrier gas and heat column to 330°C for 4 h;
3. cool and condition at 250°C for 18 h with carrier gas flowing.

Meprobamate had a retention time of 5 min and the internal standard, 12 min. Meprobamate recovery ranged from 99.5 to 101.5%. The method was also applied to commercial tablets: dose stated, 400 mg; dose found, 387.6 ± 1.4 mg. Analysis of bulk powder gave a recovery of 99.48 ± 0.42%. The authors stated that the use of tybamate would permit the analysis of related propanediol dicarbonates such as carisoprodol and mebutamate.

Interestingly, in the same year, Martis and Levy (1972) described the decomposition of meprobamate while carrying out the GC method given in the USP XVIII. The decomposition took place in the injection port, and the authors proposed that meprobamate was first hydrolyzed to 2-methyl-2-propyl-1,3-propanediol and then silylated:

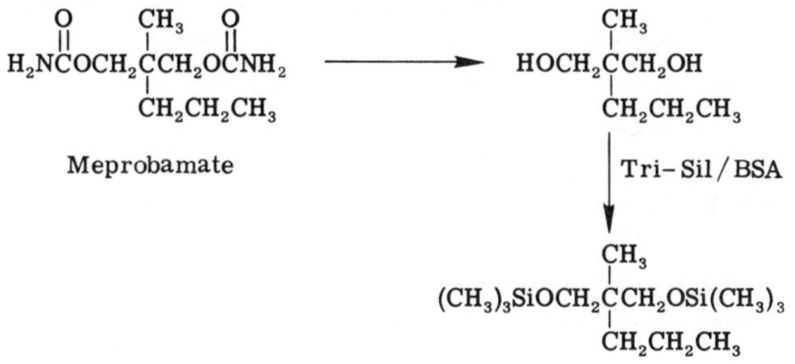

$$H_2NC(=O)OCH_2C(CH_3)(CH_2CH_2CH_3)CH_2OC(=O)NH_2 \longrightarrow HOCH_2C(CH_3)(CH_2CH_2CH_3)CH_2OH$$

Meprobamate

$$\xrightarrow{\text{Tri-Sil/BSA}} (CH_3)_3SiOCH_2C(CH_3)(CH_2CH_2CH_3)CH_2OSi(CH_3)_3$$

No mention of decomposition was made by Rabinowitz and colleagues.

Table 5.5

Relative Retention Times of Phenothiazines[a]

Compound	Relative retention time (min)	
	5% SE-30	10% DC-200
Phenothiazine	0.43	0.52
Promethazine	1.00	1.00
Diethazine	1.37	1.10
Oxomemazine	1.60	1.25
Profenamine	1.80	1.35
Chlorpromazine	2.00	1.50
Levopromazine	2.20	1.60

[a] From Mestres and Berges (1970).

A study of phenothiazines in tablets, aqueous solutions, creams, and suppositories has been published by the French workers Mestres and Berges (1970). Two columns of different polarity were used: 5% SE-30 and 10% DC-200, both operated at 210°C. The formulation to be tested was simply dissolved in a methanolic solution of the internal standard, dibutyl phthalate. The retention times relative to promethazine are given in Table 5.5.

Three years after this, there appeared a very detailed study of phenothiazines and related drugs (De Leenheer, 1973). A wide range of liquid phases was examined: SE-30, OV-1, OV-17, Lexan, STAP, QF-1, XE-60, FFAP, Versamid 900, and Carbowax 20 M. For qualitative work the preferred phases were 5% OV-1 and 2.5% FFAP plus 5% KOH on silanized Chromosorb W (100–120 mesh) or Aeropak 30 (100–120 mesh). Glass columns (1.8 m × 3 mm) were used, operated isothermally at temperatures between 180 and 280°C. Nitrogen at a flow rate of 25–30 ml/min was used as the carrier gas with flame ionization detection. The retention times, relative to imipramine, and plate numbers are given in Table 5.6.

There appears to be little work on the measurement of benzodiazepines in drug formulations, using gas chromatography, except for the measurement of diazepam and oxazepam in tablets (De Meijer, 1973). To measure diazepam, a tablet containing the drug was powdered and then ultrasonicated for 15 min with a solution of madazepam in acetone. After centrifugation, a portion was injected onto a column packed with 3% OV-17 on Gas Chrom Q (80–100 mesh) operated at a temperature of 270°C with nitrogen at 30 ml/min as the carrier gas. A similar technique was used for oxazepam, except that diazepam was used as the internal standard. The precision of the method appeared to be good; the coefficient of variation for 10 determinations ranged from 1.2 to 6.2% for diazepam and 1.8 to 5.0% for oxazepam.

IV. Stimulants

Methamphetamine hydrochloride has been measured in sustained-release tablets (Senello, 1971). To produce this sustained-release formulation, the drug was added to a methyl acrylate–methyl methacrylate copolymer, which was then compressed and treated to give the desired rate of release. Tablets produced by this procedure are extremely hard and insoluble in most solvents. The standard method of leaching out the active drug with water is a very lengthy procedure, and instead, Senello dissolved the complete tablet in chloroform and liberated methamphetamine free base by the addition of potassium hydroxide. A portion of the resulting solution was then injected directly onto a column of 5% OV-101 maintained at a temperature of 100°C. Analysis could be completed within 1 h of receiving the sample.

De Fabrizio (1972) attempted the analysis of amphetamine and amobarbital in timed-release capsules but found that the separation achieved was not good enough to provide a reliable method of estimation. However, Tammilehto *et al.* (1982) studied the decomposition of doxepin solutions exposed to daylight and radiation from a mercury lamp. Solutions containing doxepin were made alkaline and extracted with chloroform, and a portion was injected onto an OV-101 capillary column (20 m × 0.35 mm ID) operated at 210°C. Decomposition at different hydrogen ion concentrations and changes in cis–trans isomer ratios could be monitored. The purity of amphetamine sulfate has been examined by gas chromatography, not to verify its purity but to identify its origin. This was done by examining traces of impurities in the raw amphetamine preparation. It was hoped that by identifying the origin of batches, a picture of the patterns of distribution could be established (Strömberg, 1975). The raw amphetamine sulfate was dissolved in water and shaken with benzene to extract trace components, which were then chromatographed on a 3% OV-17 column using an effluent splitter and both flame ionization and electron-capture detection. Enantiomorphs of amphetamine and related amines have been resolved by gas chromatography (Beckett and Testa, 1972).

V. Steroids and Glycosides

Although an immense amount of information has been published on the gas chromatography of steroids and glycosides of all types, very little is related to their measurement in pharmaceutical preparations. This is a pity in view of the potential of gas chromatography in this type of analysis.

One of the earliest attempts was made by Talmadge *et al.* (1965) who measured ethynylestradiol in tablets or oil. A quantity of powdered tablet

Table 5.6

Relative Retentions, r_{21}, and Plate Numbers, N, of Some Representative Phenothiazines and Related Drugs with Respect to Imipramine[a]

Substance	5% OV-1		5% OV-17		5% Lexan		5% STAP	
	r_{21}	N	r_{21}	N	r_{21}	N	r_{21}	N
Aminoalkyl phenothiazines								
Dialkylaminoethyl derivatives								
Diethazine	1.62	2,676	1.76	2,773			1.99	3,07
Profenamine	1.60	2,948	1.59	2,909			1.55	3,27
Promethazine	1.17	3,410	1.32	3,280	1.61	242	1.55	3,15
Dialkylaminopropyl derivatives								
Alimemazine	1.27	3,603	1.33	3,553			1.45	3,15
Aminopromazine	1.96	2,949	2.09	3,037			2.20	3,27
Levomepromazine					—	—		
Promazine	1.37	3,443	1.60	2,934	2.07	365	2.02	3,33
Triflupromazine	1.03	3,387	0.86	3,034			0.93	3,01
Piperidylalkyl phenothiazines								
Propericiazine							—	—
Piperazinylalkyl phenothiazines								
Dixyrazine							—	—
Thioproperazine							—	—
Trifluoperazine							4.63	2,87
Azaphenothiazines								
Isothipendyl	1.18	3,441	1.36	3,361			1.57	3,24
Prothipendyl	1.45	3,494	1.71	3,078			2.10	3,19
Dibenzazepines								
Iminodibenzyl derivatives								
Desmethylimipramine	1.07	5,717	—	—	—	—	—	—
Imipramine	1.00	3,700	1.00	3,006	1.00	187	1.00	3,08
Trimeprimine	1.00	3,202	0.92	2,855			0.82	2,72
Dibenzodiazepines								
Dibenzepine								
Dibenzocycloheptadienes								
Amitriptyline	0.94	4,431	0.91	3,710	0.81	160	0.82	3,14
Nortriptyline	0.99	4,182	1.08	3,653	—	—	—	—
Retention time, t_R (min), of imipramine	4.8		11.4		12.6		12.7	

[a] From De Leenheer (1973). Nine different columns (1.80 m × 3 mm) were used. All liquid phases packe

5% QP-1		5% XE-60		5% KOH + 2.5% FFAP		5% KOH + 2.5% Versamid 900		5% KOH + 2.5% Carbowax 20 M	
r_{21}	N	r_{21}	N	r_{21}	N	r_{21}	N	r_{21}	N
		1.93	1,832	1.97	2,711	2.01	1,509		
		1.65	1,520	1.58	2,773	1.66	1,158		
1.34	3,091	1.46	2,180	1.48	2,725	1.44	1,880	1.54	3,766
1.37	2,527	1.39	2,315	1.40	2,791	1.38	1,925	1.44	3,661
		2.11	1,986	2.17	2,780	2.12	1,176		
2.86	3,611			3.65	2,718	3.21	1,881	3.91	3,835
1.66	3,578	1.81	2,277	1.92	2,741	1.81	1,407		
1.32	2,693	1.13	1,859	0.97	3,006	0.90	1,678	0.97	3,463
				—	—	—	—		
				—	—	—	—		
				—	—	—	—		
4.51	2,877			5.02	2,310	4.52	1,851	4.80	3,593
1.15	2,771	1.25	2,230	1.46	2,204	1.38	1,834	1.51	3,525
1.51	3,419	1.62	2,347	1.99	2,483	1.87	1,951	1.99	3,623
1.23	2,676	1.40	2,862	1.45	2,804	1.37	1,843	1.51	3,705
1.00	2,094	1.00	2,212	1.00	2,875	1.00	1,603	1.00	3,310
		0.85	2,043	0.82	2,791	0.84	827		
4.47	2,964			4.77	2,954	4.65	2,389	4.92	3,978
0.84	1,569	0.85	1,598	0.84	2,851	0.87	1,882	0.85	3,441
1.11	1,907	1.15	2,750	1.19	2,780	1.12	1,807	1.21	3,456
3.0		2.3		5.1		10.3		13.4	

on Aeropak-30, 100–120 mesh. Reproduced with permission from Elsevier Science Publishers.

or oil, equivalent to about 0.5 mg of ethynylestradiol, was shaken with 50 ml of heptane and extracted with 3 × 10 ml 10% sodium hydroxide. The combined aqueous extracts were acidified with sulfuric acid and extracted with 3 × 25 ml chloroform. The chloroform was washed with 15 ml water and evaporated. The internal standard, 0.05% estrone in 1 ml acetone was then added, and the solution was again evaporated. Conversion to the acetyl derivatives was carried out by the addition of 1 ml acetic anhydride and 0.2 ml pyridine and heating to 70°C for 30 min. The solution was then evaporated, and the residue was dissolved in 0.5 ml carbon disulfide; 1–2 μl of this was injected onto a column packed with a mixture of 4% SE-30 and 0.2% Epon 1001 on silanized Chromosorb P. In nine assays of dosage forms, the mean recovery was 100.1% with a standard deviation of 2.7%.

A comparison of the USP procedure with gas chromatography for the quantitative determination of ethynylestradiol was published in the following year (Boughton *et al.*, 1966). Using a simplification of the USP procedure, ethynylestradiol was extracted from tablets and granules and converted to the trimethylsilyl ether before being run on a 6-ft × 0.25-in column of 3.8% SE-30 on Anakrom ABS at 260°C. Estrone was again used as the internal standard and 0.1 mg of ethynylestradiol could be detected with a coefficient of variation of 8.6%. No interference from dimethisterone was observed. The average coefficient of variation of the USP procedure was rather better at 5.3%.

Gas chromatography has been used to measure relatively low levels of steroids in ovulation control formulations (France and Knox, 1966). The tablets were weighed, powdered, and dissolved in 4 ml water. The steroids were then extracted with 2 × 2 ml of dichloromethane after adding cholestane or progresterone as internal standard. The solvent was evaporated, the residue redissolved in 300 μl of dichloromethane, and a portion injected onto a 4-ft × 4-mm glass column packed with 3.5% SE-30 on Diatoport S (80–100 mesh) at 223°C. The retention times are given in Table 5.7. The method was applied to tablets stated to contain 0.075 mg of ethynylestradiol and a combination of ethynylestradiol (0.1 mg) and megestrol acetate (4.0 mg).

Estradiol monoesters in oil solutions have been measured by GLC after preliminary separation by TLC (Moretti *et al.*, 1969). The esters measured were estradiol 17β-cyclopentylpropionate and β-estradiol-3-benzoate at concentrations of 2 mg/ml. The steroids were converted to their TMS ethers, using a mixture of pyridine, hexamethyldisilazone, and trimethylchlorosilane (9 : 3 : 1). The GLC was carried out on a 2.2-m × 2.5-mm glass column packed with 2% SE-30 at 190°C for the cyclopentylpropionate derivative and 235°C for the benzoate derivative. Recoveries were greater than 99%. The same group of workers applied this method to measure anabolic, androgenic, estrogenic, and progesteronic steroids (Cavina *et al.*,

Table 5.7

Retention Times of Steroids in Ovulation Control Formulations[a]

Compound	t_R (min)
Cholestane	20.5
Progesterone	17.2
Ethynylestradiol	12.8
Megestrol acetate	43.0
17α-Ethynylestradiol 3-methyl ether	10.8
Norethynodrel	10.1

[a] From France and Knox (1966). Reproduced by permission of Preston Publications, Inc.

1970). In order to simplify the procedure, TLC was replaced by extraction, the steroids being partitioned between hexane and 85% ethanol. With the exception of estradiol 17β-valerate, the steroids were chromatographed without derivatization. The chromatographic details are given in Table 5.8.

Oestrone in skin creams and lotions has been measured by GLC using equilenin as internal standard (Karkhanis and Anfinsen, 1970). The cream or lotion was heated with 10% sodium hydroxide and, after filtration and pH adjustment to between 9 and 9.5, was extracted into chloroform. After concentration, a portion was injected directly onto a 3% OV-1 column operated at 245°C. Mean recoveries were good (98.9%) and precision was high (CV 1.1%).

A group of workers in France have published a detailed study of a number of steroids in injectable preparations. The steroids were extracted using acetonitrile (Mestres *et al.*, 1972) or preparative TLC (Youssef and Mestres, 1973), and 5% SE-30 or 5% OV-17 columns were used. Results are given for progesterone and the propionate, isocaproate, phenylpropionate, and deconoate esters of testosterone.

A simple extraction with 87% ethanol has been described for the quantitative determination of testosterone propionate in oily solutions (*Injectabile testosteroni*) (Holch, 1972). The paper is a detailed study of the best extraction conditions and internal standard. Gas chromatography was carried out on a 1.5-m × 4-mm glass column packed with 3.8% OV-1 on Diatoport S (80–100 mesh) at a column temperature of 230–250°C. The method was accurate, and results of the analysis of 1 and 5% *Injectabile testosteroni* were given.

The sulfate conjugates of estrogen have also been measured by GLC (McErlane and Curran, 1977). Sulfatase enzyme was used to split the conjugates, followed by chloroform extraction of the estrogens. Derivatization to the silyl ethers was carried out using a mixture of pyridine and Tri-Sil TBT (2 : 1) or 2% methoxamine in pyridine, followed by the silylating reagent.

Table 5.8

Operating Conditions for Gas Chromatographic Analysis[a]

Steroid	Column	Derivative	Column temperature (°C)	Internal standard	R_M value	Retention time of the steroid (min)	Retention time of the standard (min)
Estradiol-17β-valerate	N. 1 : 3% JXR on silanized 100–120 mesh Gas-Chrom P; 2.20 m length	Mono TMSE	230	5 α-cholestane 3 β-ol-acetate (200 μg)	1.559 ± 0.032 (±2.0%) (5)	11	17
Testosterone-17β-propionate	N. 2 : 3% QF-1 on 100–120 mesh Gas-Chrom Q 1.80 m length	None	240	estrone-3-benzoate (200 μg)	0.897 ± 0.035 (±3.9%) (6)	8	21
Progesterone	N. 3 : 1% OV-17 on 100–120 mesh Gas-Chrom Q 2.20 m length	None	250	estradiol dipropionate (150 μg)	1.092 ± 0.26 (±2.4%) (6)	8	13
Testosterone-17β-cyclopentyl propionate	N. 2	None	240	estrone-3-benzoate (250 μg)	0.928 ± 0.034 (± 3.66) (6)	30	21
19-Nortestosterone-17β-phenylpropionate	N. 1	None	240	5 α-cholestane-3β-ol-acetate (30 μg)	0.264 ± 0.008 (±3.0%) (6)	20	9
19-Nortestosterone-17β-phenylpropionate	N. 3	None	270	—	—	22	—
19-Nortestosterone-17β-decanoate	N. 1	None	240	5 α-cholestane-3β-ol-acetate (30 μg)	—	20	9
19-Nortestosterone-17β-decanoate	N. 3	None	270	—	—	15	—
19-Nortestosterone-17β-propionate	N. 1	None	210	5 α-cholestane-3β-ol-acetate (150 μg)	—	9	33
19-Nortestosterone-17β-propionate	N. 3	None	270		—	3	—

After completion of derivatization, a sample was chromatographed on a 6-ft × 0.25-in glass column packed with 3% OV-225 on Chromosorb W (HP) (100–120 mesh) operated at about 225°C. Chromatograms illustrating separations achieved with both the trimethylsilyl and methoxamine–trimethylsilyl derivatives are given in Fig. 5.2. Results of the analysis of 1.25-mg, 2.5-mg, and 25-mg tablets were presented, and those for the 1.25-mg tablet are given in Table 5.9. The authors discussed the use of OV-225 compared to diethylene glycol succinate and concluded that although the latter gave slightly better resolution, OV-225 appeared to be better for long-term use because of its stability and robustness.

Finally, the rapid quantitative analysis of digoxin in tablets has been reported by Kibbe and Aranjo (1973). An amount of crushed tablets equivalent to 0.1 mg digoxin was heated with 2 ml 50% pyridine and 0.1 ml 0.1 *N* sodium hydroxide for 30 min on a steam bath to convert digoxin to digoxigenin. The solution was evaporated to dryness and dissolved in anhydrous pyridine (100 μl), and 10 μl 0.2% cholesterol in pyridine was added as

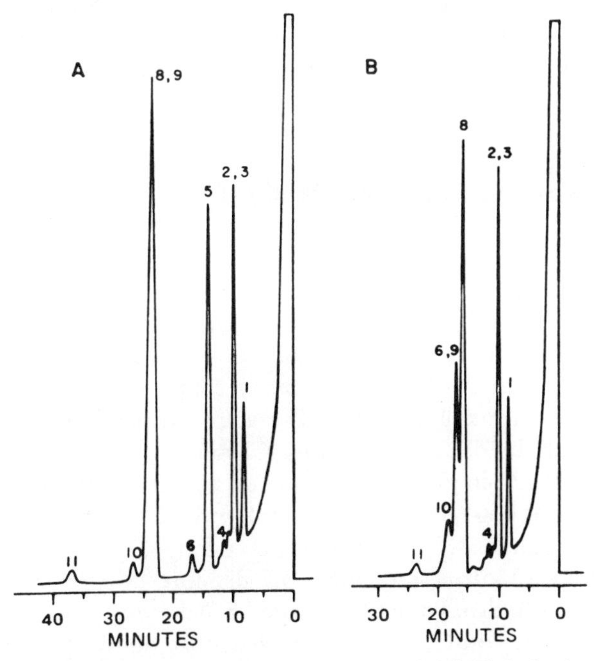

FIG. 5.2. Trimethylsilyl (A) and methoxamine-trimethylsilyl (B) derivatives of equine estrogens obtained from a 2.5-mg commercial tablet. Key: 1, estradiol-17α; 2, estradiol-17β; 3, dihydroequilin-17α; 4, dihydroequilin-17β; 5, ethinyl estradiol; 6, dihydroequilenin-17α; 7, dihydroequilenin-17β; 8, estrone; 9, equilin; 10, 8-dehydroestrone; and 11, equilenin. From McErlane and Curran (1977). Reproduced with permission of the copyright owner.

Table 5.9

Assay of 1.25-mg Tablet[a]

Steroid	Weight (mg)	Percentage
Estradiol-17α	0.08	5.5
Estradiol-17β–dihydro-equilin-17α	0.21	14.5
Dihydroequilin-17α	0.01	0.7
Dihydroequilin-17β	0.03	1.7
Estrone	0.01	0.3
Equilin	0.74	50.6
8-Dehydroestrone	0.32	21.9
Equilenin	0.03	2.0
	0.04	2.7
	1.46 mg	(116% of label claim)

[a] From McErlane and Curran (1977). Reproduced with permission of the copyright owner.

internal standard. One μl was injected onto a 2-m × 2-mm glass column packed with 2.5% OV-1 or 3.5% OV-17 on Chromosorb A (80–100 mesh) at 285°C. The retention time of digoxigenin was less than 5 min, and no derivatization was needed.

VI. Antibiotics and Other Antiinfective Drugs

The gas chromatography of halogenated hydroxyquinoline drugs has been described by Gruber *et al.* (1972). In the paper, derivatization of the sample (10 mg) with 1 ml of *N*-trimethylsilylimidazole for 15 min at room temperature was reported. The use of halogenated hydroxyquinolines in oral formulations is now banned in many countries, but some forms for local application are still available. The active principle was extracted by organic solvents before derivatization. Chromatography was carried out, using a 6-ft × 0.125-in glass column packed with 3% OV-1 on Varaport 30 (80–100 mesh) at 230°C with helium at 40 ml/min as the carrier gas. The following compounds were successfully chromatographed: 8-hydroxyquinoline, 5-chloro-8-hydroxyquinoline, 5,7-dichloro-8-hydroxyquinoline, 5-chloro-7-iodo-8-hydroxyquinoline, and 5,7-diiodo-8-hydroxyquinoline. The accuracy was found to be better than that of IR spectroscopy, and sensitivity was 5–20 times greater. The method was also claimed to be excellent for detecting impurities in halogenated quinolines.

Resorcinol monoacetate determination in dermatological products has been reported, the monoacetate being converted to the diacetate with acetic anhydride in pyridine before gas chromatography (Karkhanis *et al.*, 1973). Orcinol was used as internal standard, and chromatography was carried out using a 5% XE-60 column at 170°C. The method was found to be both rapid and reproducible, with an average recovery of 99.8%.

The gas chromatographic properties of ethambutol and other antituberculous drugs have been investigated (Richard *et al.*, 1974). No attempt, however, appears to have been made by the authors to measure the drugs in pharmaceutical formulations. Silylation was carried out using trimethylsilylimidazole, and chromatography was carried out on a column of 3% OV-17 on Chromosorb W HP (100–120 mesh) at 150°C. The retention times for the derivatives of isoniazid, ethambutol, and pyridoxine were 1.8, 3.8, and 7.2 min, respectively. The peaks, however, appeared to be rather broad. In a more recent study, Ng (1982) assayed ethambutol by taking 0.05 g of powdered tablet, adding 0.5 ml BSTFA and 0.5 ml chloroform, and heating for 30 min. The mixture was then evaporated to dryness and redissolved in 5 ml of chloroform, and a portion was chromatographed on a 3% OV-17 column, Gas Chrom Q (100–120 mesh) support, at 150°C. Unusually, he used a mass spectrometer as his detector! However, this method should work with a more modest flame ionization detection (FID) system.

Temperature-programed GLC has allowed the assay of phenol, chloroxylenol, and lidocaine hydrochloride in topical antiseptic cream (Palermo and Lundberg, 1978). The sample (6 g) was extracted directly with dimethyl sulfoxide (10 ml), which contained the internal standards (*p*-cresol, 4-chlorophenol, and 2-amino-4-phenylthiazole for phenol, chloroxylenol, and lidocaine, respectively). *N,N*-Dimethylformamide (10 ml) was then added, and the mixture was shaken until completely homogenized. More *N,N*-dimethylformamide (10 ml) was added, and, after centrifugation, the solution was diluted to 50 ml with the same solvent. A 1-μl volume was injected onto the chromatograph column, which was silanized stainless steel (1.8 m × 3 mm), packed with 3% OV-225 on Supelcoport (80–100 mesh) with helium at 30 ml/min as the carrier gas. The column was held at 90°C for 2 min, then programed to 225°C at 20°C/min, and held there for a further 4 min. Good resolution of all the constituents was obtained (Fig. 5.3). Recoveries were high, the average being 100.6% for phenol, 100.3% for chloroxylenol, and 99.1% for lidocaine hydrochloride. The authors maintained that separate internal standards were needed for each compound because of the temperature programing. About 8 samples could be assayed each working day, a significant saving in time over the existing procedure.

Before antibiotics can be chromatographed by GLC, conversion to more volatile derivatives is necessary; this is usually done by using silylating agents. Margosis (1968) describes a method for the determination of lincomycin. Earlier methods, which involved silylation in pyridine, followed by

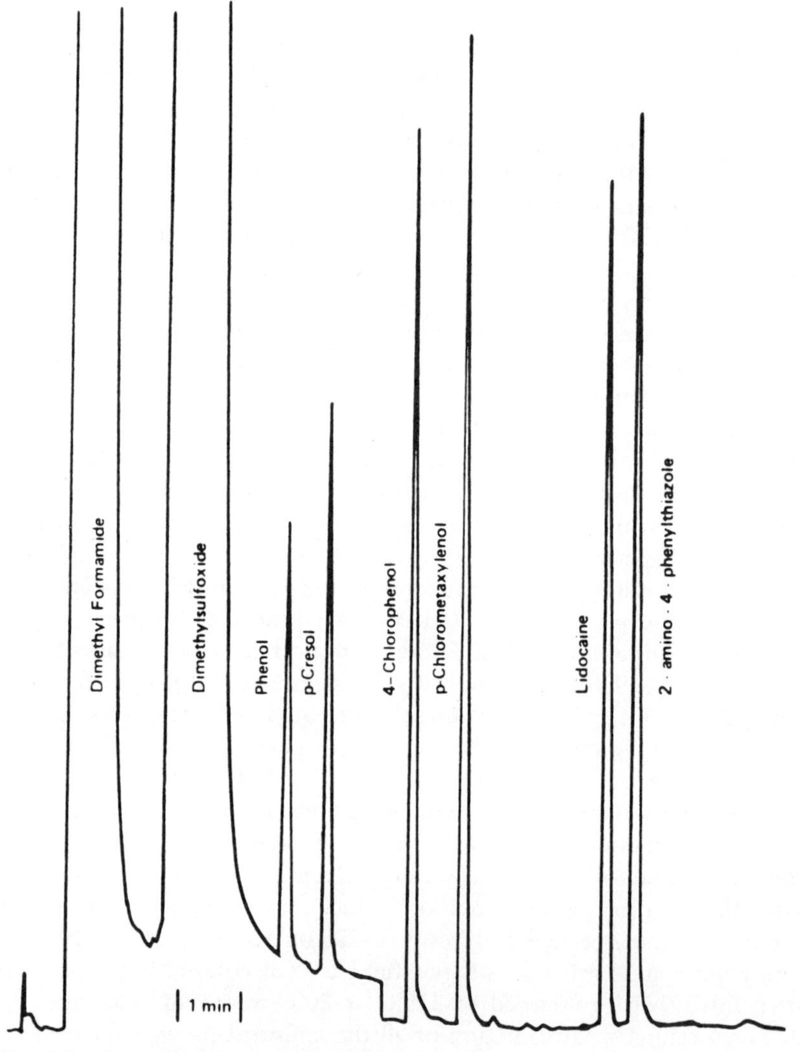

FIG. 5.3. Chromatogram of sample solution of antiseptic agents. From Palermo and Lundberg (1978). Reproduced with permission of the copyright owner

direct injection, produced rapid contamination of the detector, which had to be removed and cleaned after every four or five injections. In Margosis's method, lincomycin was dissolved in pyridine and the silylating reagent was added (filtered hexamethyldisilazane–trimethylchlorosilane, 9 : 1). The mixture was allowed to stand for not less than 30 min before the internal standard solution was added. The internal standard was a saturated solution of tetraphenylcyclopentadienone in cyclohexane. After this, water was added, the mixture shaken, and a portion of the cyclohexane phase was injected. Chromatography was carried out on a 6-ft × 3-mm glass column packed with 5% SE-30 on Gas Chrom Q (80–100 mesh) at 257°C. Nitrogen at a high flow rate (150 ml/min) was used and flame ionization was the means of detection. Under these conditions, the retention times of lincomycin B, lincomycin, and the internal standard were 6.1, 7.65, and 11.1 min, respectively. The method compared well with the microbiological assay and was applied to bulk materials, injectables, capsules, syrups, and sensitivity powders. The sucrose present in lincomycin syrup was found, initially, to cause some interference, but most of it could be precipitated by the addition of absolute ethanol. As an alternative, the lincomycin could first be extracted with chloroform from an alkaline solution and then derivatized.

The assay of chloramphenicol in bulk material, capsules, tablets, solutions, and ointments has also been described by Margosis (1970). The drug was converted to a silyl derivative, using bis(trimethylsilyl)acetamide in acetronitrile with *m*-phenylene dibenzoate as internal standard. Chromatography was carried out on a 6-ft × 3-mm glass column packed with 5% DC-200 on Gas Chrom Q (80–100 mesh) operated at 240°C with nitrogen as the carrier gas at 50 ml/min. The retention time of chloroamphenicol under these conditions was 9.5 min, and the retention times of 16 other related compounds were given. The GLC method was found to differentiate between chloramphenicol and the other structurally related compounds. A few years later, Janssen and Vanderhaeghe (1973) were able to demonstrate that the silylating agent used by Margosis, bistrimethylsilylacetamide, gave a mixture instead of a single derivative. They found that only by using a mixture of hexamethyldisilazane and trimethylchlorosilane in pyridine could a single derivative be produced. Their detailed study illustrates the need for great care in choosing the correct reaction conditions and reagents when more than one reactive group is present.

An interesting collaborative study has been described where 17 laboratories in the United States, Canada, England, Belgium, Sweden, Germany, and Denmark performed the same chloramphenicol assay (Margosis, 1974). Each laboratory received the same protocol for GLC analysis, three "unknown" samples, two samples of bulk material, and the internal standard, *m*-phenylene dibenzoate. Derivatization was carried out using bis(trimethylsilyl)acetamide in acetonitrile. Several of the investigators reported one or two extraneous peaks, which were ascribed to the mono- and tristrimethylsilyl derivatives (see the work of Janssen and Vanderhaeghe discussed in the preceding paragraph). This was corroborated by using an OV-17 column, which afforded greater resolution than the OV-1 column. The mean recoveries were very good, ranging from 100.04 to 101.37%, and mean coefficients of variation ranged from 2.16 in bulk material to 3.68% in the capsule granulation. It was concluded that the superiority of the method had been demonstrated, and its inclusion into the Code of Federal Regulations was recommended.

A rapid method for checking griseofulvin in capsules, tablets, suspensions, and bulk material has been documented (Margosis, 1972). To each dried sample, 1 ml internal standard solution was added (5 mg tetraphenylcyclopentadienone in 1 ml chloroform), and, after shaking, a portion of the supernatant was injected onto a 3-ft × 4-mm column packed with 1% OV-17 on Gas Chrom Q (100–120 mesh) operated at 225°C with helium at 60 ml/min as the carrier gas. Under these conditions, the retention time of griseofulvin was 12 min. An impurity in the reference material was found and identified by GC–MS as dechlorogriseofulvin, its retention time being 7.2 min. The gas chromatographic method was found to compare well with the official method. The impurity (dechlorogiseofulvin) was found to range from less than 0.15% to 1.99% in the pharmaceutical formulations but rose as high as 8.85% in the bulk material.

Neomycin in petroleum-based ointments has been measured by gas chromatography (Van Giessen and Tsuji, 1971). A quantity of ointment weighing about 5 g (equivalent to approximately 25 mg of neomycin sulfate) was heated for 3 min at 60°C with 25 ml chloroform. After vigorous shaking, the solution was centrifuged and the chloroform was discarded. The aqueous residue was washed with more chloroform, which was also discarded. The residue was then dissolved in 5 ml of water and shaken with 15 ml of heptane. After centrifugation, the heptane was discarded and 1 ml of the aqueous phase was freeze-dried. This residue was heated at 75°C for 35 min with 50 μl of trimethylsilyldiethylamine and trilaurin (2 mg as internal standard) in Tri-Sil Z (1 ml). A portion of this reaction mixture was injected onto a glass column (2 ft × 3 mm) packed with 3% OV-1 on Gas Chrom Q (100–120 mesh) operated at 290°C with helium at 70 ml/min as the carrier gas. Mean recoveries were high (98–100%), and the coefficient of variation was low (2%). Bacitracin, polymyxin, cortisone acetate, hydrocortisone, prednisolone, methylprednisolone, and fluoromethalone did not interfere.

Tetracyclines have also been chromatographed as their trimethylsilyl ethers (Tsuji and Robertson, 1973). A quantity of powder equivalent to 10 mg tetracycline was treated for 24 h at room temperature with 1 ml of a mixture of pyridine, trimethylsilylacetamide, and trimethylchlorosilane (10 : 5 : 5) containing 1 μg of trioctanoin per ml as internal standard. A portion of this mixture was injected directly onto a glass column (1.85 m × 3 mm) packed with 3% JXR on Gas Chrom Q (100–120 mesh), maintained at 260°C with helium at 55 ml/min as the carrier gas. The method was applied to powders, and the results agreed closely with the bioassay and spectrophotometry. The coefficient of variation of the gas chromatographic method was 2.28%.

Miribel *et al.* (1983) used gas chromatography to measure chlorhexidine in pharmaceutical formulations. Earlier methods had hydrolyzed the drug to *p*-chloroaniline, using a strong solution of sodium hydroxide. The *p*-chloroaniline was then converted to *p*-chloroiodobenzene for chromatography with electron capture detection. This compound has been claimed to be a degradation product of chlorhexidine itself. The method of Miribel and colleagues converts the drug to its trimethylsilyl derivative using BSA (room temperature, 2 h) or BSTFA (80°C, 90 min).

VII. Alkaloids and Other Related Bases

Alkaloids and related compounds can be measured directly by gas chromatography with or without derivatization. Occasionally they are measured indirectly. For example, homatropine methyl bromide has been determined in tablets and elixirs after hydrolysis to mandelic acid:

$$
\begin{array}{l}
H_2C—CH—CH_2 \quad\quad O \\
\quad |\quad\; {}^{+}| \quad\quad | \quad\quad\;\; \| \\
\quad |\; CH_3NCH_3\; CHO—C—CHC_6H_5 \\
\quad |\quad\quad | \quad\quad | \quad\quad\quad\;\; | \\
H_2C—CH—CH_2 \quad\quad\; OH \\
\quad\quad Br^{-}
\end{array}
\longrightarrow
\begin{array}{l}
H_2C—CH—CH_2 \\
\quad |\quad\; {}^{+}| \quad\quad | \\
\quad |\; CH_3NCH_3\; CHOH \\
\quad |\quad\quad | \quad\quad | \\
H_2C—CH—CH_2 \\
\quad\quad Br^{-}
\end{array}
+
\begin{array}{l}
\quad\quad\; O \\
\quad\quad\; \| \\
HO—C—CHC_6H_5 \\
\quad\quad\quad\;\; | \\
\quad\quad\quad OH \\
\text{Mandelic acid}
\end{array}
$$

Homotropine methyl bromide

The mandelic acid was then silylated before chromatography (Grabowski *et al.*, 1973).

Briggs and Simons (1983) have measured atropine in formulations that also contain cholinesterase-reactivating oximes and their degradation products. The preparation containing the atropine was saturated with ammonium sulfate, made alkaline, and extracted with methylene chloride. The organic phase was evaporated, the residue dissolved in fresh methylene chloride, and a portion was chromatographed on a 3% OV-17 column (Gas Chrom Q 80–100 mesh) operated at 190°C. Mepyramine maleate was chosen as internal standard, and detection was by flame ionization. This method allows atropine to be determined over the range 0.14–1.14 mg/ml and may well prove useful since the authors claim that the spectrophotometric method is unsatisfactory when partially degraded oximes are present.

The gas chromatographic properties of quinine, quinidine, and possible contaminants have been reported (Smith *et al.*, 1973). Thin-layer and gas–liquid chromatography were both used, but no single system was found capable of separating all of the compounds studied. The trimethylsilyl derivatives of the following were chromatographed: cinchonidine, cinchonine, dihydrocinchonidine, dihydrocinchonine, epiquinidine, epiquinine, dihydroquinidine, dihydroquinine, quinidine, quinine, quinindine, quinotoxine, and the thioglycerol adduct of quinidine. Interestingly, the dihydro analogs of quinidine and quinine were found in all of the 75 samples analyzed.

The quantitative analysis of the major alkaloids in raw gum opium has been the subject of a study by Furmanec (1974). The gum opium was cooled to −10°C and powdered. One gram of this was then shaken with 30 ml water, 1 g isoascorbic acid, and 50 mg sodium hydrosulfite for 2 h. The mixture was filtered, and the filtrate was washed with 0.25 *M* aqueous isoascorbic acid. Ten grams of sodium chloride was added, and the solution was extracted with 50 ml chloroform–isopropanol (85 : 15). The mixture was then heated to 50°C, the pH was adjusted to 8.8 with 10% ammonium hydroxide, and the solution was extracted with more chloroform–isopropanol. The organic phase was evaporated, and the residue was transferred, using methanol–chloroform (25 : 75), to a flask containing 50 mg didecyl phthalate and 25 mg resmethrin as internal standards. A portion of this solution was injected onto a 6-ft × 0.08-in glass column packed with a 1 : 1 mixture of 3% OV-17 and 5% SE-30 on Varaport 30 (80–100 mesh) and Chromosorb W (80–100 mesh AW, DMCS), respectively. The column was maintained at 250°C for 5 min following injection, then increased to 280°C at 48°C/min. Helium at 30 ml/min was used as the carrier gas. Under these conditions the retention times were as given in Table 5.10. Both internal standards were claimed to be excellent; the average percentage ± SD of morphine was 9.47% ± 0.17 (resmethrin) and 9.64 ± 0.18 (didecyl phthalate).

A thorough study of the GLC measurement of 18 basic nitrogenous drugs in a number of different preparations was published by Greenwood and Guppy (1974). These authors used a 1-m × 4-mm glass column packed with 3% OV-17 on Gas Chrom Q (80–100 mesh) with argon at 50 ml/min as the carrier gas. The drugs studied were amethocaine hydrochloride, atropine sulfate, butacaine sulfate, cinchocaine hydrochloride, cocaine hydrochloride, codeine phosphate, eserine sulfate, homatropine hydrobromide, hyoscine hydrobromide, lignocaine hydrochloride monohydrate, morphine sulfate, oxybuprocaine hydrochloride, papaverine hydrochloride, papaverine sulfate, pethidine hydrochloride, pilocarpine hydrochloride, procaine hydrochloride, and quinidine sulfate. The column temperatures ranged from 205°C (lignocaine hydrochloride monohydrate) to 270°C (papaverine salts and quinidine sulfate). A number of compounds were used as internal standards: butacaine sulfate, cinchocaine hydrochloride,

Table 5.10

Retention Times of Some Alkaloids[a]

Compound	t_R (min)
Resmethrin	2.72
Codeine	3.51
Morphine	4.25
Thebaine	7.63
Papaverine	9.15
Didecyl phthalate	9.80
Narcotine	15.65

[a] From Furmanec (1974). Reproduced with permission of Elsevier Science Publishers.

chlorpheniramine maleate, diphenhydramine hydrochloride, and procaine hydrochloride. Several of the drugs decomposed on column to give two or more peaks, e.g., atropine sulfate, eserine sulfate, and hyoscine hydrobromide. The authors claimed that accurate information could be obtained by using the summed areas of the decomposition peaks. A variety of pharmaceutical preparations were analyzed including eye drops, injection and topical solutions, sprays, and pastilles. The average recovery ranged from 96.0 to 104.0%, and coefficients of variation were low.

Moore (1983) produced a method for detecting impurities in manufactured heroin, using a fused-silica capillary column and electron capture detection. Heptafluorobutyryl derivatives were formed, and the limit of detection was in the low nano- and high femtogram range.

VIII. Essential and Other Oils

Gas chromatography is extremely suitable for the qualitative and quantitative analysis of materials as complex as the essential oils. The oils are usually hydrolyzed and then separated into fatty acids, sterols, alphatic alcohols, di- and triterpenes, etc. These, with the exception of the fatty acids, are usually chromatographed without derivatization. The large number of individual constituents of widely differing boiling points makes temperature programing essential in many cases. Some of the early studies include investigation of orange essence (Karasek, 1969), spermaceti oil (Carlier *et al.*, 1970), maize, soybean, and sunflower oils (Delaveau and Hotellier, 1971), and almond, hazelnut, and stones oils (Hotellier and Delaveau, 1972). The composition of *Lavandula stoechas* L. has been studied, using a column of 5% Carbowax 20M over the range 80–180°C. It was found to be composed mainly of camphor and fenchone with lesser

Table 5.11

Relative Retention Times of Some Essential Oils[a,b]

Compound	Relative t_R
Column at 140°C	
α-Pinene	0.05
β-Pinene	0.065
Limonene	0.08
Cineole	0.085
Paracymol	0.095
Citral	0.61
Guaiacol	0.65
Phenol	1
Column at 160°C	
Camphor	0.35
Menthol	0.45
α-Terpineol	0.50
Citral	0.62
Guaiacol	0.65
Phenol	1
Thymol	2.10
Carvacol	2.30

[a] Conditions: stationary phase, Ucon polar; volatilizing oven temperature, 195°C; carrier gas flow (nitrogen), 10 ml/min.
[b] From Mathis and More (1970).

amounts of pinene, camphene, limonene, eucalyptol, *p*-cymene, linalool, boronyl acetate, borneol, and carvacol.

A study of the volatile ingredients of suppositories and ointments has appeared, using an original method of introducing the volatile materials onto the column (Mathis and More, 1970). Usually the fats used in suppositories and ointments display the same solubilities as the active ingredients, and some sort of distillation is necessary for a preliminary separation. The above authors, however, used a volatilizing oven (*Verflüchtigungsofen*) connected directly to the injection port of the chromatograph. The preparations were dissolved in carbon disulfide and introduced into the oven where the volatiles boiled off and were carried by a stream of nitrogen onto the column. The fats remained behind and were unable to harm the stationary phase. Two column temperatures were used depending on the volatiles present. The results are shown in Table 5.11.

The composition of pine-needle oil was investigated by first using column chromatography to separate the polar terpene alcohols, acetates, and oxidized components from the nonpolar terpenes and sesquiterpenes (Joye *et al.*, 1972). Gas chromatography was carried out, using a 15-ft ×

Table 5.12

Relative Retention Times of the Components of Pine Needle Oil[a]

Compound	t_R	Compound	t_R
Bornylene	0.89	α-Fenchol	5.15
α-Pinene	1.00	Bornyl acetate	5.38
Camphene	1.18	β-Terpineol	5.91
β-Pinene	1.29	Terpinene-4-ol	6.16
Δ^3-Carene	1.37	Caryophyllene	6.78
Limonene	1.65	α-Terpineol	7.91
β-Phellandrine	1.76	Borneol	8.37
P-Cymene	2.12	Cadinene	8.83
trans-Dihydra-α-terpineol	4.65		

[a] From Joye *et al.* (1972). Reproduced from *J. Chromatogr. Sci.* by permission of Preston Pubs., Inc.

3/16-in copper column packed with 20% Carbowax 20M on Chromosorb W (70–80 mesh) at 158°C. Helium at 100 ml/min was used as the carrier gas. In all, 27 different components were identified. Among those present were those shown in Table 5.12 (with retention times relative to α-pinene).

Separation of menthol–menthone stereoisomers has also been reported (Gillin and Scanlon, 1972). The authors evaluated several liquid phases and found that Carbowax 400 provided the best overall separation. A 10-ft × 1/8-in stainless steel column was used with a 10% coating of stationary phase on Chromosorb P AW/DMCS (80–100 mesh). The column was maintained at 150°C, and helium at 60 ml/min was used as the carrier gas. Menthone, isomenthone, neomenthol, neoisomenthol, menthol, and isomenthol were resolved.

The determination of 1,8-cineole in oils of cardamom, rosemary, sage, and spike lavender has been described, using 10% PEG 400 on Chromosorb W AW (85–100 mesh) at 90°C (Analytical Methods Committee, 1973), and the same group recommended a method for the measurement of geraniol in oils of citronella. A column packed with 10% Carbowax 20M on Chromosorb W AW DMCS (80–100 mesh) was used at 140°C.

Conder *et al.* (1983) have examined the thermal decomposition of PEG 20M and some essential oils during gas chromatography and the effects of oxygen on the process. After a few hundred hours of analyzing essential oil samples at 170°C, their PEG 20M column had darkened in color and had a strong smell of acetic acid. In this interesting and detailed study, the authors showed that oxygen and the acidic products of oxidation could act as catalysts for the nonoxidative decomposition of the poly(ethylene glycol) stationary phase.

A careful study of some of the papers discussed above should provide convincing evidence of the strength of gas chromatography as a tool for the analysis of these, often very complex, mixtures.

IX. Vitamins

Gas chromatography has been applied to the study of both fat- and water-soluble vitamins, but derivatization is often necessary to increase the volatility of these compounds. The power of gas chromatography is illustrated by its ability to separate α-, γ-, and, δ-tocopherols (as their acetate derivatives) in natural oils (Hotellier and Delaveau, 1972). The determination of pantothenates and panthenol in various pharmaceutical preparations has been described (Tarli *et al.*, 1971). The panthothenates and panthenol are hydrolyzed by acid, and the resulting pantoyl lactone was chromatographed on a 1% Carbowax 20M column. The concentration of pantothenate found ranged from 93.8 to 99.0% of the label claim. Niacinamide and lidocaine could also be measured on the same column. This proved extremely useful since the latter two substances were often found together with pantothenates in pharmaceutical preparations.

Another example of an indirect estimation of a vitamin is the measurement of nicotinamide after conversion to nicotinonitrile (Vessman and Strömberg, 1975). Trifluoroacetic acid was used to dehydrate nicotinamide to nicotinonitrile, the reaction being catalyzed by pyridine:

$$\text{C}_5\text{H}_4\text{N}{-}\text{C}(=\text{O}){-}\text{NH}_2 \xrightarrow{\text{CF}_3\text{CO}_2\text{H, Py}} \text{C}_5\text{H}_4\text{N}{-}\text{C}{\equiv}\text{N}$$

Chromatography was carried out on a 1.5-m × 0.18-cm glass column packed with 20% Carbowax 20M on Gas Chrom P (100–120 mesh, acid washed and silanized) at 155°C. 6-Methylnicotinamide was used as a standard, and the method was applied to multivitamin preparations. One tablet was extracted with an acidic buffer (3 ml, pH 1.2) and diluted to 25 ml with methanol. A 0.25-ml portion of this solution was mixed with 0.2 ml internal standard solution (6-methylnicotinamide, 1 mg/ml), then the solution was evaporated to dryness on a water bath, and the residue was dissolved in 0.4 ml pyridine for derivatization. Although the method was claimed to have been applied to multivitamin formulations, no detailed information of the results was given.

The gas chromatography of a number of vitamins as silyl derivatives was the subject of a publication by Janecke and Voege (1968). A 3% SE-30 column was used, starting at 100°C and rising to 270°C at 10°C/min. Nicotinic acid, nicotinic acid amide, orotic acid, *p*-aminobenzoic acid, pyridoxolium hydrochloride, and inositol were successfully separated. Pantothenol and ascorbic acid were only partially resolved from each other. An attempt was made to separate the following from a multivitamin capsule: nicotinic acid amide, pyridoxine, ascorbic acid, vitamin A palimitate, vitamin E acetate, and rutin. There were, however, a number of interfering peaks.

Although gas chromatography can be useful for the derivatization of certain stable vitamins, many, frankly, can be analyzed more readily by high performance liquid chromatography. A rich source of information is the book by Hashmi (1973).

X. Local Anesthetics

Local anesthetics are relatively simple to measure by gas chromatography and are usually injected directly as the salts. In the heated injection port they are converted to the free bases. An extensive study has been published by Büchi *et al.* (1972). Nine different series were examined (Table 5.13). A number of stationary phases were considered: 3% SE-30, 3% JXR, 3% XE-60, 3% Apiezon-L, 3% QF-1, 3% Hi-Eff-8BP, and 2% Versamid-900. Column temperatures ranged from 180 to 240°C. Tables of retention indices are also given by the authors.

Table 5.13

Structures of Anesthetics Studied by Büchi *et al.* (1972)

Local anesthetics		
1. Anesthesin series	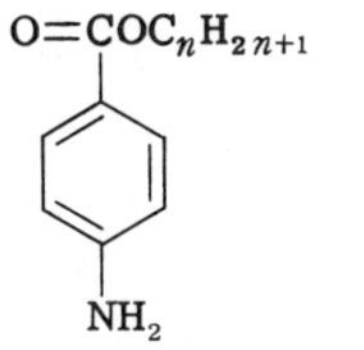 $O{=}COC_nH_{2n+1}$ on benzene ring, para NH_2	n = 1–8
2. Procaine series A	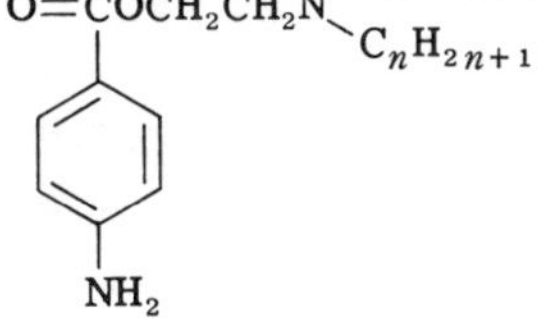 $O{=}COCH_2CH_2N(C_nH_{2n+1})_2$ on benzene ring, para NH_2	n = 1–4

(Continued)

Table 5.13 (*Continued*)

Local anesthetics		
3. Procaine series B	$O{=}CO(CH_2)_n{-}N(C_2H_5)_2$ on a benzene ring, para-NH_2	n = 2–4
4. Tetracaine series	$O{=}COCH_2CH_2N(CH_3)_2$ on a benzene ring, para-NHC_nH_{2n+1}	n = 1–6
5. Stadacaine series	$O{=}COCH_2CH_2N(C_2H_5)_2$ on a benzene ring, para-OC_nH_{2n+1}	n = 1–8
6. Cinchocaine series	$O{=}CNHCH_2CH_2N(C_2H_5)_2$ on a quinoline ring (N), with OC_nH_{2n+1}	n = 1–6
7. Isoanesthin series	$O{=}COC(R)_2{-}C(R)_2{-}R$ on a benzene ring, para-NH_2	R = H or CH_3
8. Isoprocaine series	$O{=}COC(R)_2{-}C(R)_2{-}N(C_2H_5)_2$ on a benzene ring, para-NH_2	R = H or CH_3

Table 5.13 (*Continued*)

Local anesthetics		
9. Paracaine series	$O{=}COCH_2CH_2N(C_2H_5)_2$ attached to a benzene ring bearing R in the para position	R = H, CH_3, C_3H_7, F, Cl, Br, I, NO_2, OH, SC_2H_5

References

Analytical Methods Committee. (1973). *Analyst* **98,** 616.
Beckett, A. H., and Testa, B. (1972). *J. Chromatogr.* **69,** 285.
Boughton, O. D., Bryant, R., Ludwig, W. J., and Timma, D. L. (1966). *J. Pharm. Sci.* **55,** 951.
Briggs, C. J., and Simons, K. J. (1983). *J. Chromatogr.* **257,** 132.
Büchi, J., Lorini, V., and Perlia, X. (1972). *Pharm. Acta Helv.* **47,** 65.
Carlier, A., Miet, C., Puisieux, F., and Le Hir, A. (1970). *Ann. Pharm. Fr.* **28,** 487.
Cavina, G., Moretti, G., and Siniscalchi, P. (1970). *J. Chromatogr.* **47,** 186.
Conder, J. R., Fruitwala, N. A., and Shingari, M. K. (1983). *J. Chromatogr.* **269,** 171.
Dechene, E. B., Booth, L. H., and Caughey, M. J. (1969). *J. Pharm. Pharmacol.* **21,** 678.
De Fabrizio, F. (1972). *J. Pharm. Sci.* **61,** 101.
De Fabrizio, F. (1980). *J. Pharm. Sci.* **69,** 854.
Delaveau, P., and Hotellier, F. (1971). *Ann. Pharm. Fr.* **29,** 399.
De Leenheer, A. (1973). *J. Chromatogr.* **77,** 339.
De Meijer, P. J. J. (1973). *Pharm. Weekbl. Ned.* **108,** 849.
France, J. T., and Knox, B. S. (1966). *J. Gas Chromatogr.* **4,** 173.
Furmanec, D. (1974). *J. Chromatogr.* **89,** 76.
Gillen, D. G., and Scanlon, J. T. (1972). *J. Chromatogr. Sci.* **10,** 729.
Grabowski, B. F., Softly, B. J., Chang, B. L., and Haney, W. G. Jr., (1973). *J. Pharm. Sci.* **62,** 807.
Greenwood, N. D., and Guppy, I. W. (1974). *Analyst* **99,** 313.
Gruber, M. P., Klein, R. W., Foxx, M. E., and Campisi, J. (1972). *J. Pharm. Sci.* **61,** 1147.
Hashmi, M. H. (1973). "Assay of Vitamins in Pharmaceutical Preparations." Wiley (Interscience), New York.
Hishita, C., and Lauback, R. G. (1969). *J. Pharm. Sci.* **58,** 745.
Holch, K. (1972). *Dansk. Tidsskr. Farm.* **46,** 169.
Hotellier, F., and Delaveau, P. (1972). *Ann. Pharm. Fr.* **30,** 495.
Janecke, H., and Voege, H. (1968). *Naturwissenschaften* **55,** 447.
Janssen, G., and Vanderhaeghe, H. (1973). *J. Chromatogr.* **82,** 297.
Joye, N. M., Proveaux, A. T., and Lawrence, R. V. (1972). *J. Chromatogr. Sci.* **10,** 590.
Karasek, F. W. (1969). *Res. Dev.* **20,** 74.
Karkhanis, P. P., and Anfinsen, J. R. (1970). *J. Pharm. Sci.* **59,** 535.
Karkhanis, P. P., Edlund, D. O., and Anfinsen, J. R. (1973). *J. Pharm. Sci.* **62,** 804.
Kibbe, A. H., and Aranjo, O. E. (1973). *J. Pharm. Sci.* **62,** 1702.

Laik Ali, S. (1974). *Chromatographia* **7,** 655.
Margosis, M. (1968). *J. Chromatogr.* **37,** 46.
Margosis, M. (1970). *J. Chromatogr.* **47,** 341.
Margosis, M. (1972). *J. Chromatogr.* **70,** 73.
Margosis, M. (1974). *J. Pharm. Sci.* **67,** 435.
Mario, E., and Meehan, L. G. (1970). *J. Pharm. Sci.* **59,** 538.
Martis, L., and Levy, R. H. (1972). *J. Pharm. Sci.* **61,** 1341.
Mathis, C., and More, M. F. (1970). *Arch. Pharm. Berl.* **303,** 657.
McErlane, K. M., and Curran, N. M. (1977). *J. Pharm. Sci.* **66,** 523.
Mestres, R., and Berges, J-L. (1970). *Trav. Soc. Pharm. Montpell.* **30,** 69.
Mestres, R., and Berges, J.-L. (1972). *Trav. Soc. Pharm. Montpell.* **32,** 313.
Mestres, R., Youssef, A. F., and Berges, J.-L. (1972). *Trav. Soc. Pharm. Montpell.* **32,** 331.
Miribel, L., Brazier, J. L., Comet, F., and Lecompte, D. (1983). *J. Chromatogr.* **268,** 321.
Moore, J. M. (1983). *J. Chromatogr.* **281,** 355.
Moretti, G., Cavina, G., and Sardi de Valverde, J. (1969). *J. Chromatogr.* **40,** 410.
Neckopulos, A. A. (1971). *J. Chromatogr. Sci.* **9,** 173.
Ng, T. L. (1982). *J. Chromatogr. Sci.* **20,** 479.
Palermo, P. J., and Lundberg, J. B. (1978). *J. Pharm. Sci.* **67,** 1627.
Patel, S., Perrin, J. H., and Windheuser, J. J. (1972). *J. Pharm. Sci.* **61,** 1794.
Rabinowitz, M. P., Reisberg, P., and Bodin, J. I. (1972). *J. Pharm. Sci.* **61,** 1974.
Rader, B. R. (1969). *J. Pharm. Sci.* **58,** 1535.
Richard, B. M., Manno, J. E., and Manno, B. R. (1974). *J. Chromatogr.* **89,** 80.
Senello, L. T. (1971). *J. Pharm. Sci.* **60,** 595.
Smith, E., Barkan, S., Ross, B., Maienthal, M., and Levine, J. (1973). *J. Pharm. Sci.* **62,** 1151.
Strömberg, L. (1975). *J. Chromatogr.* **106,** 335.
Talmadge, J. M., Penner, M. H., and Geller, M. (1965). *J. Pharm. Sci.* **54,** 1194.
Tammilehto, S., Heikkinen, L., and Järvelä, P. (1982). *J. Chromatogr.* **246,** 308.
Tarli, P., Benocci, S., and Neri, P. (1970). *Farmaco. Ed. Prat.* **25,** 504.
Tsuji, K., and Robertson, J. H. (1973). *Anal. Chem.* **45,** 2136.
Van Giessen, B., and Tsuji, K. (1971). *J. Pharm. Sci.* **60,** 1068.
Vessman, J., and Strömberg, S. (1975). *J. Pharm. Sci.* **64,** 311.
Visalli, A. J., Patel, D. M., and Reavey-Cantwell, N. H. (1976). *J. Pharm. Sci.* **65,** 1686.
Watson, J. R., Lawrence, R. C., and Lovering, E. G. (1978). *J. Pharm. Sci.* **67,** 950.
Wong, C. K., Urbigkit, J. R., Conca, N., Cohen, D. M. (1973). *J. Pharm. Sci.* **62,** 1340.
Youssef, A. F., and Mestres, R. (1973). *Trav. Soc. Pharm. Montpell.* **33,** 35.

Chapter 6

Measurement of Drugs in Body Fluids

Without doubt the field of drug analysis in body fluids has allowed gas chromatography to demonstrate most clearly its power and versatility. Indeed, the rapid development of drug analysis during the 1960s was due mainly to the appearance of the gas chromatograph as a commercial instrument. The development of more powerful, that is, more pharmacologically active drugs that can be given in lower doses has ensured the continuing use of this selective and sensitive technique. High performance liquid chromatography (HPLC) is now widely available, and it offers the possibility of simpler analytical methods because less sample "purification" may be needed. However, the applicability of HPLC has been hampered by the fact that the most commonly used detectors, UV and fluorescence, are less sensitive than electron-capture detection (ECD) or MS. Many drugs do not absorb well in the UV region or possess native fluorescence, and derivatization techniques, analogous to those of GLC, have been developed to attach UV-absorbing or fluorescent moieties to drug molecules. This approach has generally been very successful but has resulted frequently in methods just as complex as those for GLC. The HPLC technique is, however, excellent for very polar and high molecular weight compounds that would not be volatile or would decompose on the GLC column, and the technique should be seen to complement rather than compete with GLC.

In the last chapter a systematic approach was adopted toward the analysis of drugs in dosage forms. Here it is planned to present a discussion of the important principles underlying the development of a method and then to provide a series of interesting examples. For those who would prefer a more systematic approach the books and reviews by Gudzinowicz (1967), Reid (1976), Kaye (1980), and Sadee and Beelen (1980) can be recommended.

I. Choice of Body Fluid

Blood (or plasma) is the obvious fluid to study since it circulates freely throughout the body and its tissues. Many workers prefer to use plamsa (or serum) in the belief that it gives cleaner extracts. In the vast majority of cases there is no objection to using plasma or serum provided it has been demonstrated previously that most of the drug is in fact in the plasma and not the red cells. Erythrocyte levels of phenobarbital, phenytoin, salicylate, chlorthalidone, and chlorpromazine have been shown to be higher than the concentration in plasma, in fact chlorthalidone red cell levels have been shown to be 50 to 100 times greater than in plasma (Fleuren and Van Rossum, 1978). Urine is also frequently analyzed because the drug (and metabolite) levels are higher than in blood. Urine drug levels can be extremely useful in calculating a number of pharmacokinetic parameters of drug excretion. It does, however, contain many more potentially interfering compounds than blood. Less frequently saliva, cerebrospinal fluid, breast milk, and other body fluids are studied. These are usually analyzed to answer specific questions on drug distribution, and present little problem analytically.

II. Extraction of Drugs from Body Fluids

Before gas chromatography is carried out, it is necessary to have some sort of purification to remove unwanted material and concentrate the drug to be measured. It has been argued on occasion that elaborate purification procedures make redundant the separating power of the chromatographic column, but experience has shown that insufficient purification leads rapidly to a deterioration of column efficiency and detector contamination. When a derivatization step is included in the method, common sense indicates that reactions are more likely to proceed smoothly if the drug is in as pure a state as possible. There are many approaches that can be used to remove a drug from body fluids, and these have been discussed in a comprehensive review by Reid (1976). A simpler treatment of the methods used is presented here.

The idea behind any method of extraction is to remove as much as possible of the drug to be analyzed from the body fluid while leaving behind the maximum amount of endogenous material. Extraction with an organic solvent is the most widely used technique and is treated first. Less frequently, adsorption procedures have been used, and these are treated next; the section concludes with a brief description of some miscellaneous methods.

A. Extraction Using Organic Solvents

The procedure of extraction using organic solvents is based on the principle that a drug in its unionized form will be more soluble in nonpolar organic solvents than will its ionized form. Hence by adjusting the pH of the biological fluid containing the drug, the degree of ionization of the drug can be altered. The use of the technique can best be illustrated by a simple diagram representing the partition of a drug between an aqueous phase and an organic nonpolar solvent (Fig. 6.1).

Consider an acidic drug with a pK_a (dissociation constant) of 4. By definition, 50% of the drug will be ionized at pH 4 and as the pH of the biological fluid is made more acidic, more drug will exist in the unionized form:

$$R{-}C(=O)O^- + H^+ \rightleftharpoons R{-}C(=O)OH$$

From Fig. 6.1 it can be seen that as the pH decreases, more and more of the drug enters the organic phase. Commonly used nonpolar solvents are hexane, toluene, diethyl ether, dichloromethane, chloroform, or mixtures of these. For basic drugs the converse is true, and increasing the pH renders the drug less ionized and more soluble in the organic solvent. Hence it can be

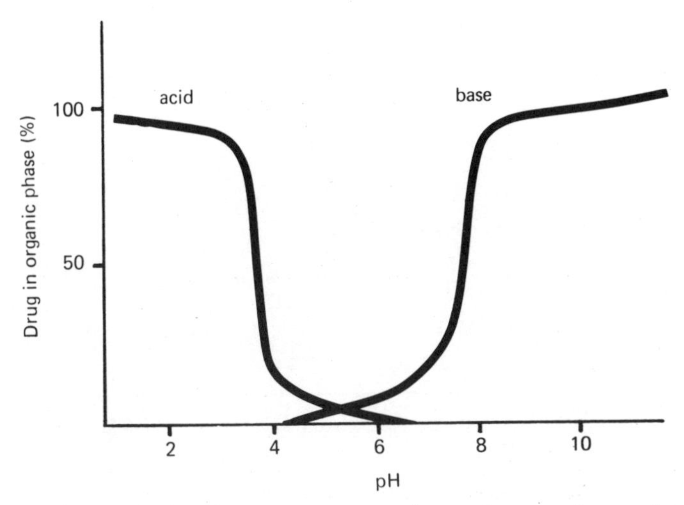

FIG. 6.1. Partition of a drug between an aqueous phase and an organic nonpolar solvent.

immediately appreciated that, by altering the pH, the separation of basic substances from acidic can be achieved. However, weak acids and bases behave somewhat differently.

Figure 6.2, taken from the work of Le Petit (1977), demonstrates the change in partition coefficient of some drugs with change in pH. Salicylic acid and the antihistamine promethazine display the characteristics of an organic acid and a base, respectively. Caffeine is a very weak base (pK_a 0.9) and phenacetin is an amide, and both show little change in partition over a wide range of pH. Both will be extracted at acidic, neutral, or basic pH. This behavior, provided it is recognized, need not present a problem.

For drugs that behave like strong acids and bases, a further purification step, called "back-extraction" can be carried out. This can be illustrated by taking as an example a basic drug of pK_a 8 present in plasma. If the plasma is brought to pH 10 using a suitable alkali or alkaline buffer and the drug is shaken with an organic solvent in which it is soluble, it will be removed from the aqueous into the organic phase. Unfortunately, many endogenous bases and neutral compounds will be extracted also if they are soluble in the organic solvent. The organic phase can now be separated from the aqueous phase (using a Pasteur pipet, for example). This is usually done after centrifugation to separate completely the two phases. By shaking the organic phase with dilute acid, such as 0.1 *M* sulfuric, the basic drug will now be

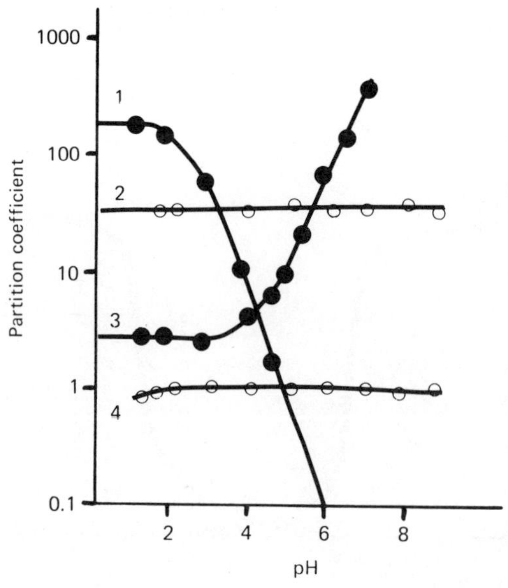

FIG. 6.2. Change in partition coefficient of some drugs with change in pH. (1) Salicylic acid, (2) phenacetin, (3) promethazine, (4) caffeine. From Le Petit (1977).

ionized and no longer be as soluble in the organic phase: it will be extracted into the aqueous phase. Any neutral compounds will be left behind in the organic phase. Hence a further purification has been achieved. Endogenous bases may also be co-extracted with the basic drug, but by carefully choosing the pH and the organic solvent the amounts of these bases can be reduced. The acidic aqueous phase can now be made alkaline by the addition of a base like 2 *M* sodium hydroxide and shaken with fresh organic solvent to take the drug into the organic phase. The organic phase can be separated, evaporated, and the next stage of the procedure carried out.

In many methods the organic phase, containing the drug, is washed with water, and the washings are discarded. This step must be carefully controlled. For example, if the drug is a moderate to strong base and the water used to wash is even slightly acidic, then much of the drug may be removed into the aqueous phase. If such a process takes place, then low and very erratic recoveries will result.

Before leaving solvent extraction a number of simple points deserve to be emphasized:

1. In general the least polar solvent capable of extracting the drug in question should be used in order to reduce the possibility of co-extracting endogenous material. The least polar are the hydrocarbons such as hexane, toluene, chlorinated hydrocarbons, and diethyl and related ethers. Ethyl acetate is more polar, and the short-chain alcohols are very polar and miscible with water to a greater or less degree.
2. Many drugs are so highly lipid soluble that even in the ionized state they can be extracted into nonpolar solvents. For example, the β-adrenoceptor blocking drug propranolol has a pK_a of about 9.5, and therefore in a pH 7.4 buffer it is more than 99% ionized. Since its partition coefficient between *n*-octanol and pH 7.4 buffer at 37°C is 20.2, this means that the ionized drug is highly soluble in *n*-octanol. This topic is treated in more detail by Le Petit (1980).
3. Many extraction procedures employ a ratio of solvent : aqueous phase greater than unity in order to reduce the possibility of emulsion formation during shaking. There are, however, successful methods that use ratios very much less than unity (see page 205). Troublesome emulsions can also be avoided or reduced by saturating the aqueous phase with an inorganic salt, such as sodium chloride, before extraction.
4. Recoveries can be increased by using mixed solvents such as hexane–butanol (9 : 1) or hexane–isoamyl alcohol (97 : 3). Such solvents are also useful in extracting polar drugs. Fluorinated alcohols have been used in solvent mixtures with success (see page 127).

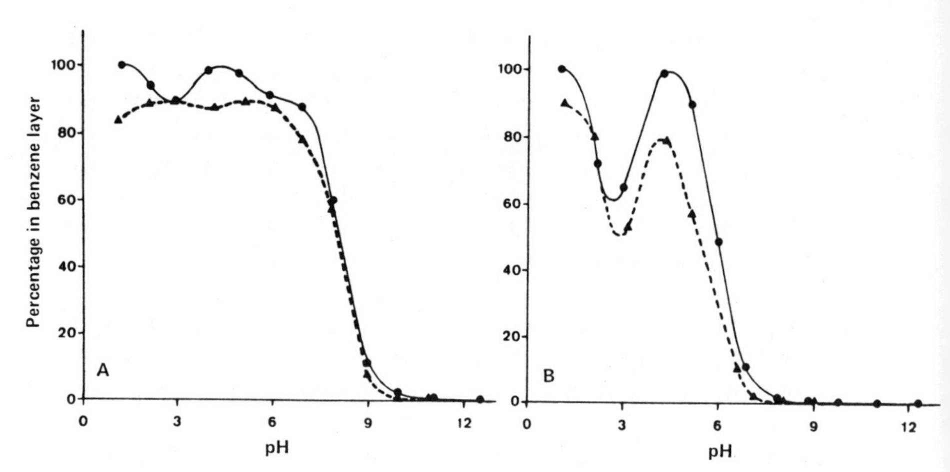

FIG. 6.3. Partition of the anti-inflammatory drug diclofenac: A, aqueous phosphate buffer; B, plasma plus phosphate buffer (1 : 3). Diclofenac sodium, ●; internal standard, ▲. From Geiger *et al.* (1975b).

5. Extraction conditions should always be optimized using the relevant biological fluid: it should not be assumed that the extractability from water will exactly mimic that from blood. To illustrate this, the partition of the lipid-soluble anti-inflammatory drug diclofenac is shown in Fig. 6.3.
6. When extraction conditions have been optimized, it is worth while putting a series of specimens through the complete procedure, half of them diluted approximately 10-fold. Although this does not always succeed, it can result in cleaner extracts.

B. Ion-Pair Extraction

This is a refinement of simple solvent extraction in which the drug Q is paired with a counterion X to give a species with an overall charge of zero and which is soluble in nonpolar organic solvents:

$$Q^{\pm}_{aq} + X^{\mp}_{aq} \rightarrow [Q^{\pm}X^{\mp}]^{0}_{org}$$

This type of extraction is particularly useful with highly polar drugs that cannot be extracted directly into organic solvents. Ion-pair systems can be characterized by an extraction constant E_{QX}, and by varying the choice of counterion and extracting solvent, the extraction can be made very selective indeed. This is illustrated in Fig. 6.4, which shows the extraction of ion pairs of nortriptyline, amitriptyline, and *N*-methylamitriptyline.

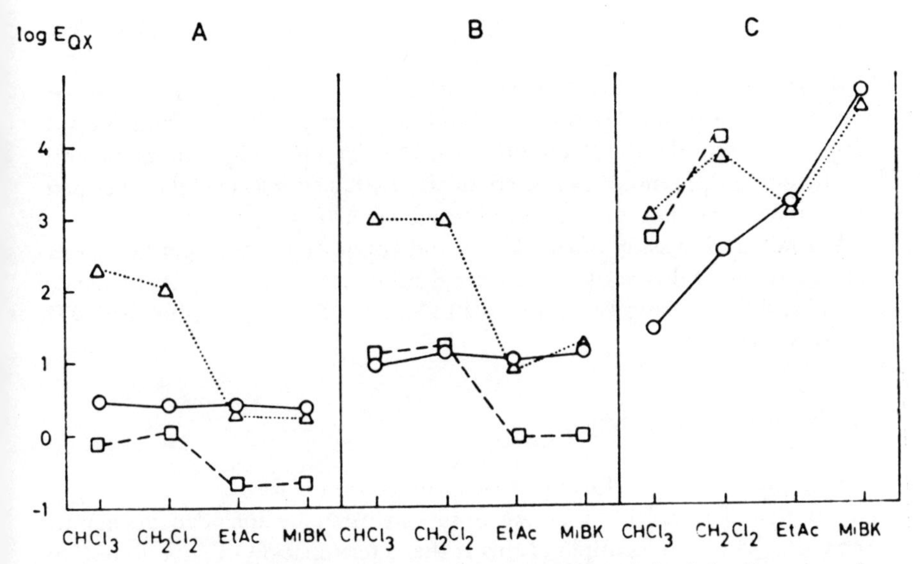

FIG. 6.4. Extraction of nortriptyline, amitriptyline, and *N*-methylamitriptyline as ion pairs into different organic solvents. A, chloride ion pairs; B, bromide ion pairs; C, perchlorate ion pairs. EtAc, ethyl acetate; MiBK, methylisobutyl ketone; ○, nortriptyline; △, amitriptyline; □, *N*-methylamitriptyline. From Modin and Schill (1975) with permission of Pergamon Press Ltd.

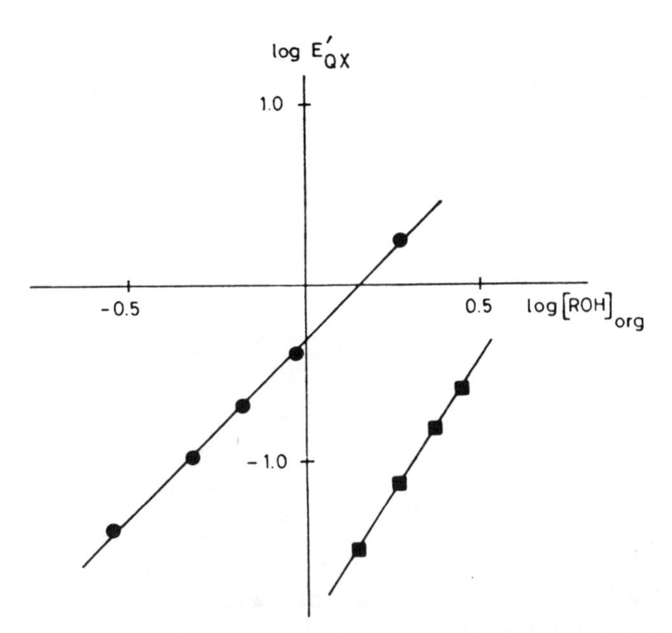

FIG. 6.5. Extraction constants of alprenolol and 4-hydroxyalprenolol with cyclohexane–1-pentanol and perchlorate as counterion (●, alprenolol; ■, 4-hydroxyalprenolol). From Borg *et al*. (1974).

Because the procedure can be made highly selective, it has also been applied to drugs that are not polar and are readily soluble in organic solvents. For example, the β-adrenoceptor blocker alprenolol and its metabolite 4-hydroxyalprenolol can be efficiently extracted into cyclohexane–pentanol, using perchlorate as the counterion (Fig. 6.5).

A detailed discussion of the theory and application of ion pairing is outside the scope of this book, but a careful study of the early work of Gustavii (1967) and the reviews by Fransson and Schill (1975) and Schill *et al.* (1977) is recommended.

C. Extractive Alkylation

A major advantage of using ion-pair extraction before gas chromatography is that derivatization of the drug can often be included during the extraction step. An example of this is the determination of clioquinol (Degen *et al.*, 1976a). The drug is amphoteric and is extracted into dichloromethane, using tetrahexylammonium as the counterion:

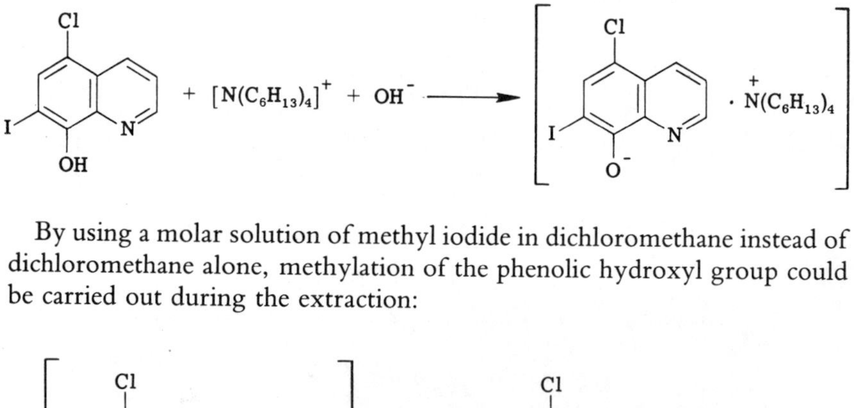

Cl, I, N, OH $+ [N(C_6H_{13})_4]^+ + OH^- \longrightarrow$ [Cl, I, N, O^- $\cdot \overset{+}{N}(C_6H_{13})_4$]

By using a molar solution of methyl iodide in dichloromethane instead of dichloromethane alone, methylation of the phenolic hydroxyl group could be carried out during the extraction:

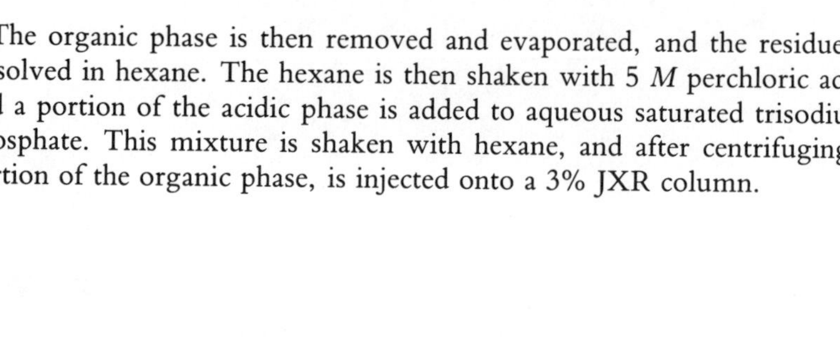

[Cl, I, N, O^- $\cdot \overset{+}{N}(C_6H_{13})_4$] $\xrightarrow[CH_2Cl_2]{CH_3I}$ Cl, I, N, OCH_3 $+ I^- \overset{+}{N}(C_6H_{13})_4$

The organic phase is then removed and evaporated, and the residue is dissolved in hexane. The hexane is then shaken with 5 *M* perchloric acid, and a portion of the acidic phase is added to aqueous saturated trisodium phosphate. This mixture is shaken with hexane, and after centrifuging a portion of the organic phase, is injected onto a 3% JXR column.

Extractive alkylation has been used to measure the diuretic chlorthalidone in plasma at very low concentrations (Ervik and Gustavii, 1974), and it has been applied to another diuretic, fenquizone, by Marzo *et al.* (1983). These

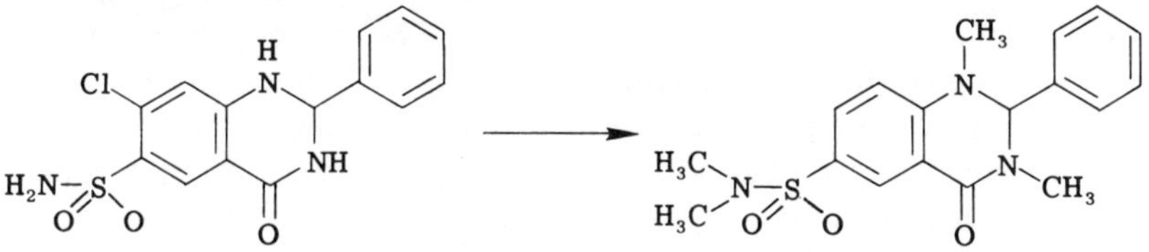

Fenquizone

latter authors used methyl iodide in dichloromethane with the addition of 0.05 ml 0.1 *M* tetrahexylammonium acid sulfate. The anticonvulsant suclofenide has also been measured by this technique (Degen and Schweizer, 1977). The reaction, of course, need not be confined to methylation: ritalinic acid, the main metabolite of methyl phenidate, has been determined after extractive alkylation, which resulted in the formation of the pentafluorobenzyl ester (Van Boven and Daenens, 1979).

D. Other Extractions with Derivatization

An interesting example of extraction with derivatization is a micromethod for the determination of paracetamol in plasma or serum (Huggett *et al.*, 1981). A small quantity of plasma or serum (0.1 ml) is shaken (Vortex) with internal standard, pH 7.4 buffer, and acylating reagent (acetic anhydride–*N*-methylimidazole–chloroform 5 : 1 : 30). The paracetamol is acetylated and extracted into the chloroform phase. The tube is then centrifuged, and a portion of the chloroform phase is injected directly onto a column of 3% Apolane 87. The recovery is claimed to be about 95%.

E. Extraction by Adsorption

Under controlled conditions some finely divided materials, when added to body fluids, will adsorb to a greater or lesser extent any drug present. The finely divided material can then be separated, usually by centrifugation, and the drug leached off by a suitable solvent.

Edwards and McCredie (1967) studied the binding of a number of acidic, basic, and neutral drugs to anionic and cationic exchange resins and to charcoal. Although the authors studied the adsorption of drugs with a view to using their results to help treat cases of drug poisoning, the findings are still applicable to analytical chemistry. Among the drugs studied were barbiturates, salicylate, quinine, morphine, strychnine, chlorpromazine, glutethimide, carbromal, and meprobamate. Charcoal was found to be most effective with the neutral drugs. Dekker *et al.* (1968) studied the adsorption of a range of drugs by activated charcoal from simulated gastric juice. Amphetamine, primaquine, chlorpheniramine, colchicine, phenytoin, aspirin, and propoxyphene were efficiently absorbed. This study was carried out with a view to treating drug overdoses, so no data on the desorption from charcoal is given. The adsorption of a similar range of drugs with the addition of amphetamine, cocaine, and common benzodiazepines has been studied by Meola and Vanko (1974). Although these authors used thin-layer chromatography to detect the drugs, much useful information can be obtained from their work. The drugs were removed from the charcoal using diethyl ether or a mixture of chloroform and isopropanol.

A good example of the charcoal adsorption method is in the determination of low levels of the antispasticity drug baclofen (Degen and Riess, 1976). It is very polar, being derived from γ-aminobutyric acid, and cannot be directly extracted from biological fluids into organic solvents. The authors compared active charcoal with XAD-2 ion exchange resin and, although good recoveries were obtained from both, preferred the charcoal because it was easier to free from interfering impurities. In addition, the ion exchange resin had to be used in small columns to give good recoveries, and this is inconvenient if a large number of specimens must be processed.

Cl Cl Cl

H_2N CO_2H → H_2N $CO_2C_4H_9$ → C_3F_7CONH $CO_2C_4H_9$

Baclofen

Internal standard and 10 mg washed charcoal are added to plasma buffered at pH 10 in a glass tube, which is sealed with a screw cap. The tube is rotated gently (20 min), centrifuged, and the supernatant dilute plasma is removed. The charcoal is washed, and the adsorbed baclofen is eluted with a few milliliters of methanol–water (9 : 1). The methanol solution is then evaporated to dryness, and the drug is converted to the butyl ester and finally to the heptafluorobutyramide for GLC. The overall recovery of the method is about 50%.

This method lends itself to large numbers of specimens, and the early steps can be carried out in standard disposable screw-cap glass tubes.

Theophylline has been removed from serum using the charcoal method, and although UV spectrophotometry was used to measure it, the optimization of the extraction conditions are interesting. The overall recovery was 60–70% with a coefficient of variation of 2–3% (Plavšić, 1978). The method has also been applied successfully to the removal of the highly water-soluble antitubercular drug ethambutol from plasma and urine. The plasma or urine was buffered at pH 10 and treated as in the method for baclofen. Better recoveries were obtained, however, if the drug was eluted with 5% trifluoroacetic acid in methanol. This evaporates rapidly, and conversion of the drug to the TFA derivative is a simple matter. The overall recovery was greater than 90% (Jack, 1978).

Anyone who has used the charcoal method cannot fail to be impressed by its relative simplicity. Large numbers of specimens can be extracted simultaneously, and since disposable screw-capped tubes can be used, there is no need for scrupulously cleaning and silanizing glassware. An added bonus is that no large volumes of plasma saturated with halogenated hydrocarbons are accumulated as in many of the methods using solvent extraction. Indeed it is surprising that the method is not more widely used, considering its simplicity and cheapness.

Alumina has been used to remove compounds from biological fluids, but its use is usually restricted to adrenaline and other catecholamines (Wong *et al.*, 1973). Ion-exchange resins can also be used, but since they are generally employed as columns, they are not well suited to processing large numbers of specimens.

F. Miscellaneous Methods

Freeze-drying biological fluids and then eluting the drug into a suitable solvent is possible but is of limited usefulness because of the expense of the freeze-drying apparatus and its inability to cope with large numbers of specimens. Thin-layer chromatography can be useful as a purification step but it is time-consuming, especially with large numbers of specimens. Another method of sample handling is worth mentioning here: theophylline, caffeine, and phenobarbital have been measured in blood collected on filter paper (Brazier and Delaye, 1981). A small quantity of blood is removed from patients or volunteers and a 30-μl sample is spotted onto a piece of filter paper and allowed to dry. Large numbers of samples can be easily transported in this way and even sent by mail. For analysis, the disk is punched out into a test tube containing a pH 5.2 buffer, and the drugs are extracted in the normal way with a chloroform–isopropanol mixture (95 : 5).

III. Developing a Method

In general, the simplest method possible should be chosen; more complex procedures may yield cleaner extracts, but where large numbers of specimens must be processed, efficiency can decline. The conditions for all steps in the method should be optimized, e.g., extraction pH, solvents, time for shaking, temperature of derivatization, and effect of a catalyst. Trevor *et al.* (1972) remarked that the time for extractions reported in many methods is usually much longer than necessary and "appears to coincide strongly with the gustatorial habits of a particular laboratory." Sometimes, however, there may be a deliberate decision to shake or derivatize longer than is absolutely necessary; this may be done in order to allow time to distil fresh solvents, prepare reagents, or start another series of extractions. The time claimed for a method from start to finish should however be realistic (Jack, 1981).

IV. Use of an Internal Standard

During extraction, back-extraction and derivatization losses of drug inevitably occur; e.g., it is impossible to remove all of the organic phase from the initial extraction or derivatization may not be complete. To compensate for such losses, many analytical chemists add a known fixed amount of a suitable compound—the internal standard—to the specimens to be analyzed and also to the spiked standards. When the chromatograms are complete, the peak height (or area) of the drug is measured relative to the peak height (or area) of the internal standard and the amount of drug present in each specimen can be calculated from a calibration curve produced by running the series of spiked samples.

The internal standard should mimic as closely as possible the behavior of the drug to be determined. If a derivatization step is included in the method, then the internal standard should be capable of being derivatized in the same way. In their method for the determination of the antihypertensive drug clonidine, Edlund and Paalzow (1977) chose a very similar compound, 2[(2,4-dichlorophenyl)amino]-2-imidazoline (Fig. 6.6).

If possible, an internal standard should be selected that has a shorter retention time than the drug itself in order to keep the time for chromatography to a minimum. Drug companies synthesize a number of analogs of each successful drug, and in general, it is not difficult to find one that is suitable. Some workers prefer to use a second drug as an internal standard so that they have two methods instead of one. For example, a GC method for a number of β-adrenoceptor blocking agents may use alprenolol as the

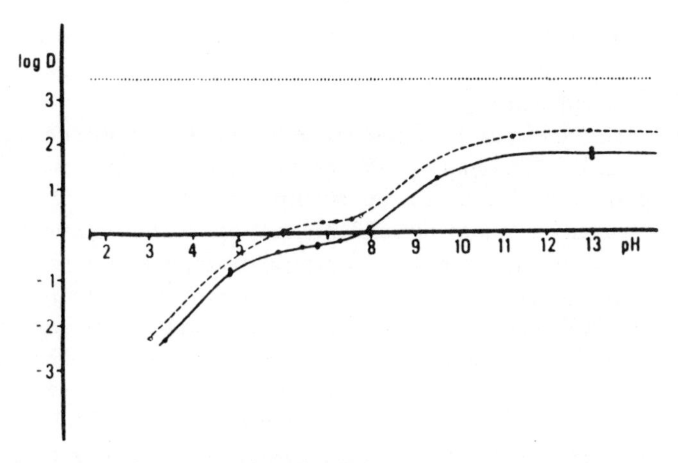

FIG. 6.6. Partition of PFB bromide and PFB derivatives between equal phase volumes of heptane and water, determined by GLC with FID detection. Solid line, PFB derivative of clonidine; broken line, PFB derivative of 2-[(2,4-dichlorophenyl)amino]-2-imidazoline; dotted line, PFB bromide. From Edlund and Paalzow (1977).

internal standard for oxprenolol, oxprenolol as the internal standard for metoprolol, metroprolol as the internal standard for propranolol, and so on. If a mass spectrometer is used as the detector, then an isotope of the drug to be measured can be used. This will, of course, mimic exactly the behavior of the drug.

Where there is no derivatization step and recoveries are high, some workers dispense with an internal standard altogether.

These then are the basic points to be kept in mind when developing a method. Let us now look in detail at a selection that will illustrate some interesting points. The author has been personally involved in the development of a number of these methods, and no apology is made for choosing them since it gives an opportunity to show what sometimes lies behind the published methods.

V. Anticonvulsants

Careful blood level monitoring of anticonvulsant drugs is necessary to reduce the incidence of seizures in epileptics. Many GC methods have been published, and, not surprisingly in view of the varied nature of the drugs, it has been found difficult to produce a single method that will give a reliable estimation of all the most commonly used drugs. In fact, many workers are moving over to HPLC or immunoassay as being simpler techniques. However, in view of the amount of effort and sheer ingenuity that has been expended on the GC of these drugs, a brief comment on some of the methods is worth while.

The methods can be grouped into those that do not involve derivatization and those that do. The main problems encountered in chromatographing underivatized anticonvulsants is the adsorption onto the column, as in the case of phenobarbital, and thermal breakdown during chromatography, as is the case with carbamazepine. A typical example of a method without derivatization is that of Heipertz *et al.* (1977) in which ethosuximide, carbamazepine, phenobarbital, primidone, phenytoin, and its metabolite phenylethylmalonediamide are extracted from serum with diethyl ether. The organic phase is removed, and ammonium sulfate and more ether is added to the serum, which is then shaken, and the precipitated proteins are centrifuged. The ether extracts are combined, dried over sodium sulfate, evaporated, redissolved in methanol, and injected onto the gas chromatograph column.

The simple extraction was claimed to be excellent for phenytoin, phenobarbital, primidone, and carbamazepine, with recoveries ranging from 94 to 100%. Ethosuximide is relatively volatile, and some loss was encountered during the evaporation of the ether at 45°C. A considerably better recovery was obtained if the ether was allowed to evaporate overnight at room temperature. The recovery of the phenytoin metabolite was 87% but seemed to be relatively constant. Under the GC conditions described by the authors, the well-established acid-catalyzed breakdown of carbamazepine in the injection port seemed to be reproducible, resulting in an iminostilbene peak equivalent to 16–22% of the carbamazepine peak. A number of stationary phases were tried in the authors' search for one that was highly polar but temperature stable. Finally, the authors chose 0.8% SP-1000 on Chromosorb 750 (100–120 mesh). Temperature programing was used, and the chromatograph was equipped with a nitrogen detector. Mephenytoin was added to the serum as the internal standard.

In the same year, Serfontein and De Villiers (1977) published a method for barbiturates and phenytoin, using trimethylanilinium hydroxide for derivatization. This method is claimed to be superior to earlier derivatization procedures that produce on chromatography an "early" phenobarbital peak due to incomplete methylation. The drugs are extracted into toluene, which is separated and vortexed with trimethylanilinium hydroxide solution. After centrifugation, the aqueous phase is removed and shaken with an acidic buffer, and a portion is injected slowly (6 sec) onto the column: 3% OV-17 on Gas Chrom Q (100–120 mesh), programed from 150 to 280°C at 15°C/min. Using this procedure there is no early phenobarbital peak.

Because of the apparent lack of correlation between the dose of sodium valproate administered and the resulting plasma levels, routine monitoring of this drug has been introduced in a number of laboratories. Sodium valproate is usually measured as the free acid after a simple extraction into a nonpolar solvent such as hexane. Fellenberg and Pollard (1977) report a recovery of only 70%, so the use of an internal standard is essential. These authors used *n*-octanoic acid with an acidic column of 10% DEGA and 2% phosphoric acid on Diatomite C (80–100 mesh) at 150°C with flame ionization detection. Sodium valproate has also been measured as the methyl (Gyllenhaal and Albinsson, 1978), butyl (Hulshoff and Roseboom, 1979), and phenacyl esters (Gupta *et al.*, 1979). An alternative approach has been tried by Cook and Jowett (1983), who converted valproic acid to its α-bromo-*p*-nitroacetophenone derivative, which was then chromatographed on a 2% OV-17 phase with flame ionization detection. The authors have claimed that their method has the advantage that a versatile phase such as OV-17 can be used in place of phases like FFAP and SP-1000, which are restricted to low-temperature operation.

Although, as mentioned earlier, HPLC and immunoassay pose strong competition to GLC for the analysis of many anticonvulsants, sodium valproate will probably continue to be measured by the latter technique. Anyone wishing a comprehensive summary of GLC methods for anticonvulsants should start with the review by Kaye (1980), in which 24 methods are listed for anticonvulsants in general, 19 for phenytoin, and 17 for sodium valproate.

VI. Antihypertensive Drugs

A. β-Adrenoceptor Antagonists

The earliest of the β-adrenoceptor antagonists, propranolol, practolol, alprenolol, and oxprenolol, quickly established a role in the treatment of hypertension. They have been joined by many others, and their role has been extended to the treatment of other diseases. Most of them can be considered as being derived from isopropylaminopropanol, they are often given in relatively low doses, and many are extensively metabolized. In order to carry out detailed pharmacokinetic studies of these drugs, sensitive and specific analytical techniques were needed. Most of the present β blockers have been assayed by GLC methods, although HPLC is being proposed as an alternative in some cases.

The majority of β blockers are readily lipid soluble, and even the most polar of them, atenolol and nadolol, are sufficiently lipid soluble to be extracted with organic solvents. All possess an alcoholic hydroxyl group and most a secondary amino group, both of which need to be converted to a less polar moiety before gas chromatography. In view of the low levels of circulating unchanged drug, electron-capture detection is necessary, and derivatization of the two polar groups is usually carried out using either trifluoroacetic anhydride or the less volatile and more expensive pentafluoropropionic anhydride, heptafluorobutyric anhydride, or its imidazole. A summary of only a selection of methods used for the most common β blockers is given in Tables 6.1 and 6.2.

There is no shortage of methods to determine β blockers. Most methods are really minor modifications of earlier ones, using a different extraction solvent or derivatizing agent. Kinney's metroprolol method shortens the retention times by having a carrier gas flow of 230 ml/min! In Jack's unpublished method (1973) for five of the most common drugs, shorter retention times were achieved by reducing both the column length and internal diameter from 2.7 m × 4 mm to 1.5 m × 2 mm. The method of Walle (1974) is much quoted, and it is apparently the only study published that reports the gas chromatography of sotalol. The present author was interested in producing a single method for the three polar β blockers atenolol, sotalol, and nadolol and tried a wide range of derivatizing agents, solvents, and catalysts in an attempt to chromatograph sotalol. Even duplicating Walle's conditions exactly failed to produce any evidence of a peak! Perhaps this might act as a challenge to someone.

Scales and Copsey (1975) measured the polar atenolol by GLC after extracting it from whole blood, plasma, or urine using *n*-butanol–cyclohexane (7:3). The volume of solvent was eight times that of the biological fluid. The atenolol was then back-extracted into 0.1 *M* HCl. This relatively polar solvent mixture was necessary to remove the drug from the biological fluid. The drug was then converted to a volatile derivative using HFBA.

$$\underset{\text{Atenolol}}{OCH_2CH(OH)CH_2NHCH(CH_3)_2\text{-}C_6H_4\text{-}CH_2CONH_2} \xrightarrow{\text{HFBA}} OCH_2CH(OCOC_3F_7)CH_2N(COC_3F_7)CH(CH_3)_2\text{-}C_6H_4\text{-}CH_2CN$$

This reagent not only acylates the alcoholic hydroxyl and the secondary amino group but also converts the phenylacetamide part of the molecule to the benzyl cyanide.

Table 6.1

Conditions for GLC of Some β-Adrenoceptor Blocking Drugs

Drug	Internal standard	Derivatizing agent	Stationary phase	Temperature (°C)	Reference
Alprenolol	Dodecachlorooctahydro-1,3,4-methano-2*H*-cyclobuto-[*c,d*]pentalene	TFAA	3% QF-1	170	Ervik (1969)
	Oxprenolol	TFAA	3% OV-17	190	Rawlins *et al.* (1974)
Atenolol	Oxprenolol	HFBA	3.8% UCW-98	185	Scales and Copsey (1975)
	H 155/87	TFAA	3% OV-1	180	Ervik *et al.* (1980)
Metoprolol	9-Bromophenanthrene	TFAA	3% JXR	160	Ervik (1975)
	9-Bromophenanthrene	PFPA	3% OV-101	195	Zak *et al.* (1980)
	Propranolol	TFAA	3% OV-17	200	Kinney (1981)
	Oxprenolol	HFBA	3% OV-1	200	Sioufi *et al.* (1983b)
Nadolol	—	HFBA	3% OV-7	200	Jack and Laugher (1982)
Oxprenolol	Alprenolol	TFAA	3% OV-101	158	Jack and Riess (1974)
	Propranolol	HFBA	3% SE-30	187	Sioufi *et al.* (1983a)
Pindolol	Propranolol	TFAIm	2% OV-17	190	Guerret *et al.* (1980)
Propranolol	Pronethalol	HFBA	3% OV-17	205	Di Salle *et al.* (1973)
	4-Methylpropranolol	PFPA	10% OV-1	245	Kates and Jones (1977)
Timolol	*N*-Isopropyl analog	HFBA	3% OV-17	165–275 (programed)	Tocco *et al.* (1975)

Table 6.2

Multiple Methods[a]

Drug combination	Derivatizing agent	Stationary phase	Temperature (°C)	Reference
Alprenolol, oxprenolol, metoprolol, propranolol, pindolol	TFAA	3% OV-101 (2 mm ID)	150	Jack (1973)
Alprenolol, oxprenolol, propranolol, practolol, sotalol	TFAA	2% OV-17	170	Walle (1974)
Oxprenolol, metoprolol, alprenolol, pindolol, propranolol	TFAA	3% JXR	200	Degen and Riess (1976)
Atenolol, metoprolol, propranolol	PFPA	3% OV-1	190	Wan *et al.* (1978)
Alprenolol, oxprenolol	HFBA	OV 101 WCOT	205 (alp) 185 (oxp)	De Bruyne *et al.* (1979)

[a] No internal standards are given, since one can be chosen from the drugs listed.

Ervik *et al.* (1980) have produced a nice study of the effect of changing the amount of sodium chloride in the aqueous phase and the percentage of alcohol in the extracting solvent on the distribution ratio of atenolol. The use of 3% heptafluorobutanol in dichloromethane extracted 92% of the atenolol when the volume of organic solvent used was eight times that of the aqueous phase (plasma or urine). Heptafluorobutanol has the disadvantage, however, of being expensive.

Perfluorinated acid anhydrides were tried unsuccessfully to derivatize pindolol, and the reproducibility was poor (Guerret *et al.*, 1980); however, a successful method was produced that uses trifluoroacetylimidazole. The authors attributed the difficulty to the indole ring and to degradation products and contamination resulting from the use of the more powerful acylating agents. The best results were obtained by using trifluoroacetylimidazole and trimethylamine in hexane. Analysis by GC–MS showed that all three labile hydrogens had been replaced:

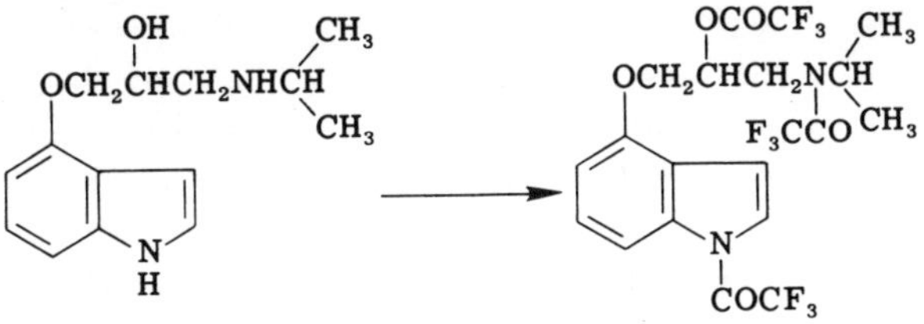

OH, CH₃, OCH₂CHCH₂NHCH, CH₃, N, H → OCOCF₃, CH₃, OCH₂CHCH₂NCH, F₃CCO, CH₃, N, COCF₃

Pindolol

Propranolol was used as the internal standard, and it was noted that practolol gave a peak with the same retention time as pindolol.

Sioufi *et al.* (1983a,b) have modified the method of Jack and Riess (1974) and measured oxprenolol and metoprolol as their heptafluorobutyryl derivatives. The derivatization reaction was carried out in hexane with pyridine as a catalyst. A novel approach to metoprolol was attempted by Gyllenhaal and Vessman (1983). The drug was extracted from alkaline plasma using a hexane–methylene chloride mixture (4 : 1), and to this organic phase was added 0.01 ml 2 *M* phosgene in toluene. This resulted in the formation of

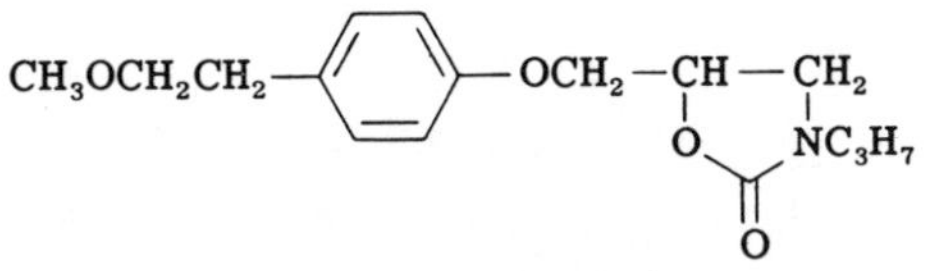

$CH_3OCH_2CH_2$–C₆H₄–OCH_2–CH–CH_2, O, NC_3H_7, O

Metoprolol oxazolidine

the oxazolidine. The organic mixture was evaporated to dryness and dissolved in ethyl acetate, and a portion was chromatographed on 3% Hi-EFF-8BP using nitrogen-selective flame ionization detection. There seems to be no sign that the flood of new β blockers is abating, and another one, betaxolol, has been measured as its HFBA derivative using electron-capture detection (Ganansia *et al.*, 1983). For the derivatization, ethyl acetate was used as the solvent, and chromatography was performed on an OV-101 capillary column (25 m × 0.2 mm ID).

B. Hydralazine and Other Hydrazines

Hydralazine and the closely related dihydralazine are powerful vasodilators used in the treatment of hypertension. These compounds are extremely reactive and can undergo many different chemical transformations. Hydralazine is more widely used therapeutically, and a great deal of work has been published concerning this extremely interesting compound. The drug is rapidly and extensively metabolized, and the persistence of its pharmacological effect long after the blood levels of the parent drug have ceased to be measurable suggests that one or more of the metabolites may be active.

The analysis of the very reactive hydralazine in body fluids presented a challenge, and since the author was closely involved in some of the early work, an account of this may be of interest. In planning the method it was known that at least two metabolites were likely to be present: *s*-triazolophthalazine **(I)** and 3-methyl-*s*-triazolo[3,4*a*]phthalazine **(II)**:

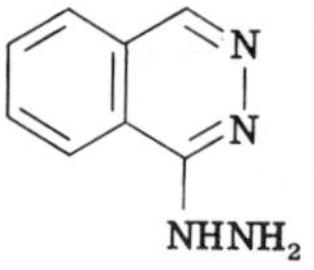

$NHNH_2$ N N N N N N N N N CH_3

Hydralazine 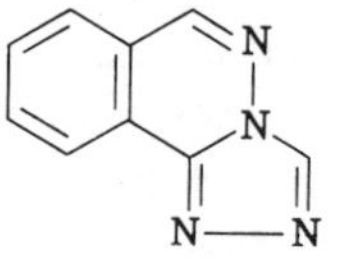 I 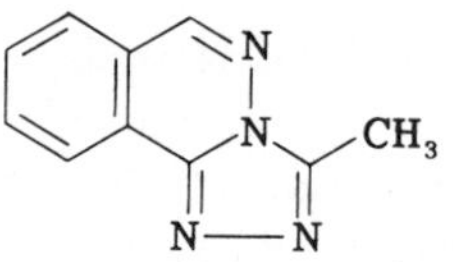 II

Thus a simple ring closure of hydralazine with formaldehyde to the *s*-triazolo derivative would be unwise. It was decided instead to use nitrous acid to give the tetrazolaphthalazine, and the 4-methyl analog was chosen as the internal standard:

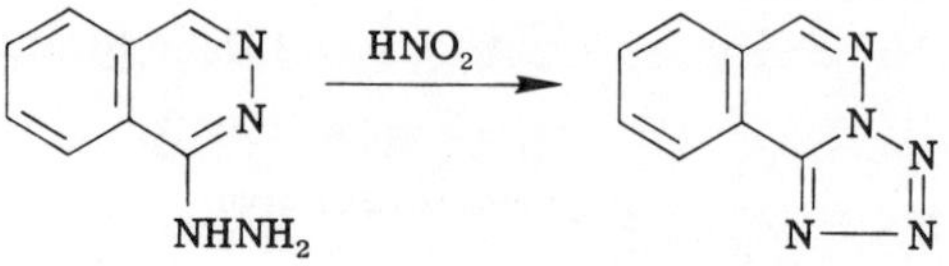

N N $NHNH_2$ HNO_2 N N N N N

Because of the known instability of hydralazine *in vitro* it was decided that derivatization should be carried out in the plasma and not following solvent extraction. This was done by adding to plasma (1 ml or less) 1 ml 2 *M* hydrochloric acid and 0.1 ml of 50% aqueous sodium nitrite.

The technique of N-FID was chosen in the hope that the metabolites could also be measured. Preliminary work showed that the polar phase OV-225 was suitable, but it soon became apparent that the detector was not sufficiently sensitive to detect the concentrations anticipated in humans after a single 50-mg oral dose. In the laboratory where the work was carried out, the gas chromatograph alongside the one equipped with N-FID had an electron-capture detector, and because of the conjugated aromatic system of the tetrazolo derivative, the author made an injection and left the laboratory to do something else. On returning he saw the following trace:

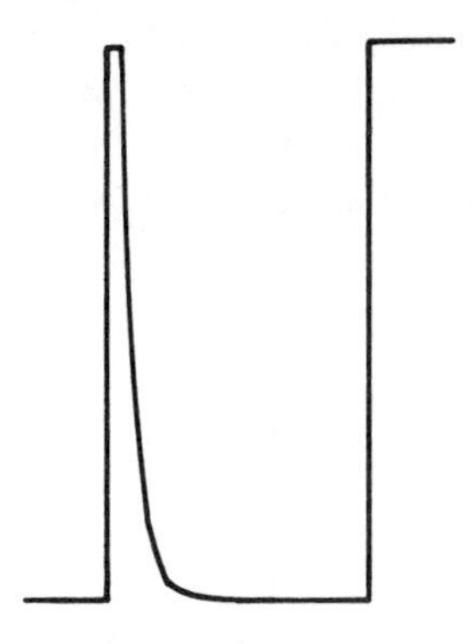

Like most analysts who had cut their teeth on gas chromatography, he had a strong pessimistic streak and thought that something had gone wrong with the flow of carrier gas. However, he decided to wait for a few moments and saw the recorder pen descend slowly only to rise again in a minute or so. He began to become slightly more optimistic—it could be that the tetrazolo derivative was strongly electron capturing and that he had just witnessed the emergence from the column of the hydralazine derivative followed closely by the internal standard. He diluted his solution 100-fold and re-injected a sample; two beautifully symmetrical peaks were obtained! The chromatograph with the N-FID was quickly stripped down, an electron-capture detector was fitted, and thus a sensitive method was produced that could detect 10-ng hydralazine per ml plasma (Jack *et al.*, 1975). No interference from the known metabolites was encountered, which unfortunately also meant that they could not be measured as had been planned.

In 1977 Zak *et al.* demonstrated that the amount of hydralazine detected in plasma depended on the acid concentration during the derivatization step and proposed that hydralazine in plasma was present at least in part as acid-hydrolyzable conjugates. Soon after (Reece *et al.*, 1978), it was shown that in humans a major metabolite, now identified as hydralazine pyruvic acid hydrazone, interfered in the GC analysis. What had been previously reported as hydralazine was actually hydralazine plus pyruvate hydrazone. Reece *et al.* described a method to allow the determination of the pyruvate hydrazone by extracting it from acidic plasma into chloroform, the extraction being carried out within 30 sec of acidification. The chloroform phase was then evaporated, and the residue was treated with trifluoroacetic anhydride at 21°C for 5 min. The contents of the tube were then evaporated to dryness, the residue redissolved in toluene, and a portion was injected onto a 3% OV-17 column operated at 190°C. 4-Methylhydralazine pyruvic acid hydrazone was added to the plasma as the internal standard.

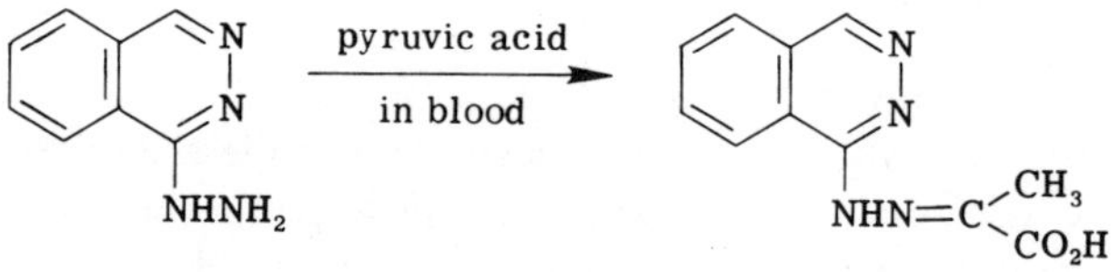

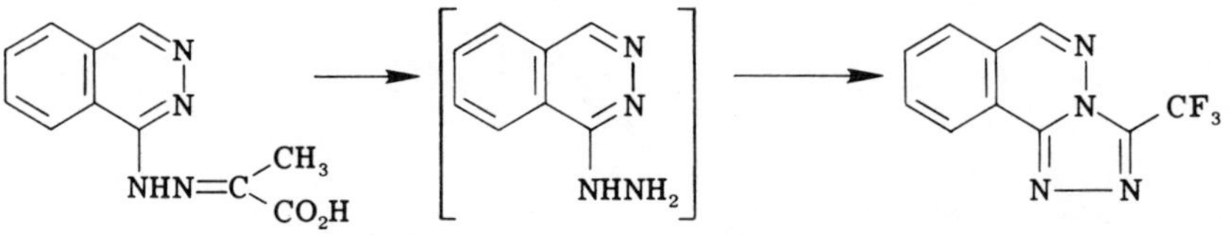

Degen reported a method of measuring unchanged hydralazine without interference from the pyruvate hydrazone (1979). This involved adding 2,4-pentanedione directly to plasma at a pH of about 6.4 and shaking at room temperature for 1 h. Free hydralazine is converted to 1-(3,5-dimethyl-1-pyrazolyl)phthalazine, which is extracted with hexane, concentrated, and chromatographed on a 3% OV-17 column at 230°C. Nitrogen-selective flame ionization was the detector chosen. Methods that determine hydralazine together with its metabolites are discussed in Chapter 7.

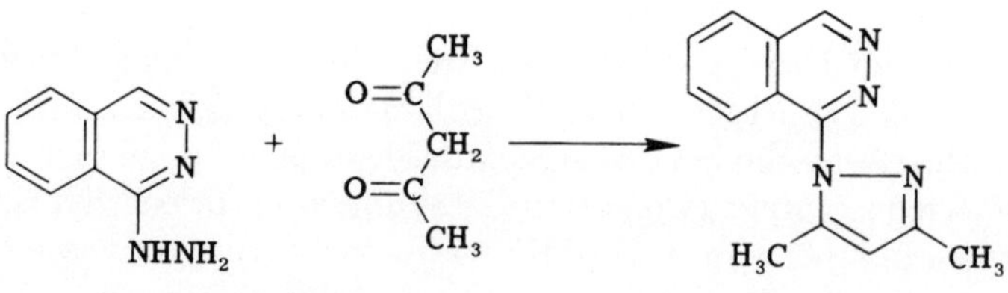

The closely related vasodilating drug dihydralazine cannot be analyzed by the nitrous acid method already described for hydralazine (Jack *et al.*, 1975) since it does not undergo conversion to the ditetrazolophthalazine but rather to an unstable azide, 5-azido-1,2,3,3*a*,4-pentazocyclopent[*e*]indene:

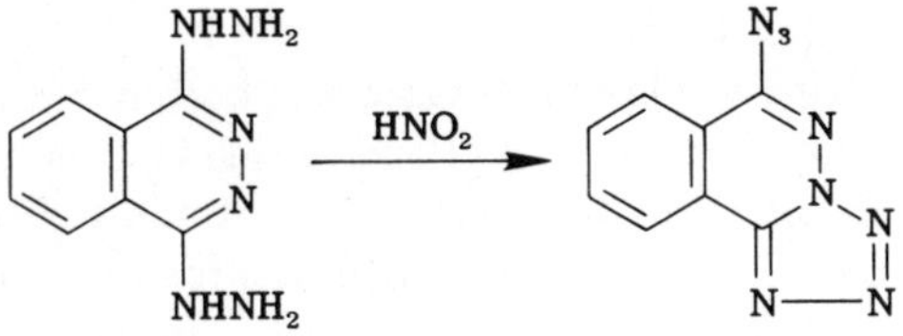

Dihydralazine

In his 1979 paper, Degen claims that 2,4-pentanedione will also react with dihydralazine; however, he and colleagues (1982) have published a method for "apparent" dihydralazine, which converts it first to the unstable azidotetrazolophthalazine, which is then treated with sodium methylate to give a more stable compound with excellent gas chromatographic properties. The derivative was chromatographed on 3% OV-225 at 250°C, and a limit of detection of 5 ng per ml plasma was reported.

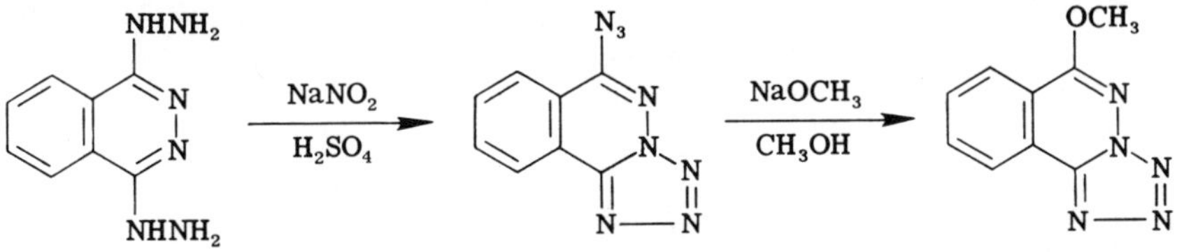

Another drug containing a hydrazine side chain is the antidepressant phenelzine, which, like hydralazine, is also unstable under basic conditions. Caddy and Stead (1977) published a method capable of measuring the drug in urine and used an interesting reaction. The sample is made strongly acidic with 50% v/v sulfuric acid and shaken with 5% potassium iodate. The

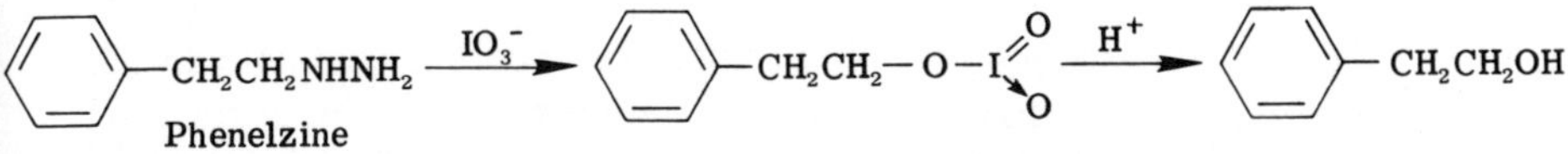

Phenelzine

hydrazone group is oxidized to 2-phenylethanol, iodine is precipitated if high concentrations of phenelzine are present, and the filtrate is extracted with diethyl ether. The ether phase is reduced in volume, and a portion is chromatographed on 10% FFAP at 150°C. When applied to urine specimens, a preliminary wash with diethyl ether under acidic conditions is carried out and the organic phase discarded; this step removes potentially interfering acidic and neutral compounds. Other pharmacologically active hydrazine derivatives such as isoniazid and nialamide would also be oxidized by iodate, but it was claimed that the oxidation products would not interfere with the measurement of 2-phenylethanol. The method was capable of detecting 0.05 μg of drug in 1 ml urine.

Cooper *et al.* (1978) produced a sensitive method to allow the measurement of phenelzine in plasma. Plasma was buffered at pH 6.8 and extracted with a mixture of benzene–ethyl acetate (4:1). The drug was then back-extracted into 0.05 *M* sulfuric acid, the pH adjusted to 6.8, and then the drug was extracted into benzene containing the internal standard 1-phenylpropylhydrazine. The benzene layer was removed, and to it was added heptafluorobutyric anhydride, and derivatization was carried out at 60°C for 30 min. The mixture was then shaken vigorously with water to which was added 1 *M* NH_4OH, followed by more shaking. The aqueous phase was discarded, and after centrifugation, a portion of the organic phase was injected onto a 5% OV-17 column at 155°C. The procedure was claimed to be capable of detecting 1 pmol of phenelzine (about 0.14 ng).

Washing the reaction mixture with weak base was believed to be essential since simply evaporating to dryness and redissolving resulted in a long solvent peak and a reduction in sensitivity. The long solvent peak most likely resulted from traces of residual heptafluorobutyric anhydride and its free acid. A reduction in sensitivity could be caused by some loss of the volatile derivative during evaporation.

C. Guanidines and Related Drugs

The guanidine group is strongly basic, and many substituted guanidines have pK_a values in excess of 13. Most of the drugs containing this group are adrenergic neurone blockers that have been used in the treatment of hypertension. One of the earliest studies concerned the drug guanoxan, which was chromatographed after conversion to the acetyl derivative, using a mixture of acetic anhydride and pyridine (Jack *et al.*, 1972). Although a symmetrical peak was obtained on a mixed phase of 7% DC-560 and 1% EGSP-Z at a column temperature of 200°C, attempts to derivatize the drug after a butanol extraction from urine were unsuccessful. Shortly after, Hengstmann *et al.* (1974) published an interesting method of measuring guanethidine and other guanidino-containing drugs in bological fluids. Initially, they had tried to derivatize the drugs, using trifluoroacetyl and pentafluoropropionyl anhydrides, but these had proved very unstable and had decomposed upon removal of the anhydride at room temperature. Accordingly, it was decided first to hydrolyze the guanidino group to the corresponding amine and then to derivatize that. Sodium, potassium, and ammonium hydroxides were all tried, and the potassium salt was found to give the best results. The hydrolysis step was carried out at three different temperatures: 90, 110, and 140°C; 110°C was found to give the maximum yield. The authors concluded that insufficient conversion took place at the lower temperature while decomposition became significant at the higher.

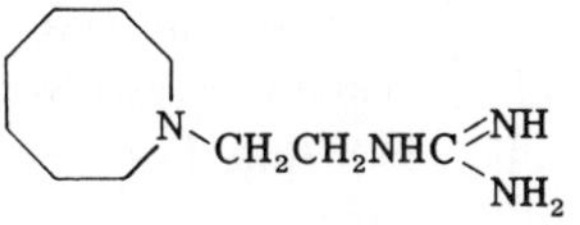

Guanethidine

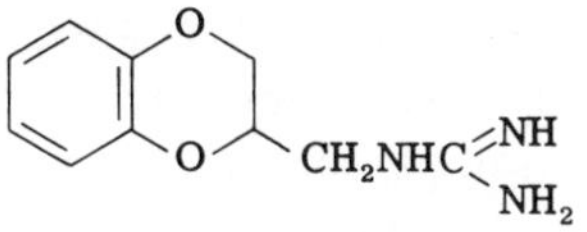

Guanoxan

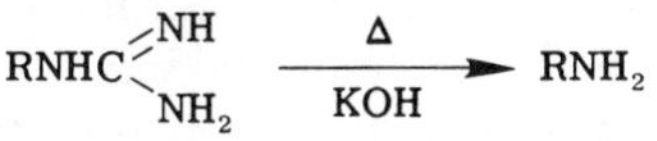

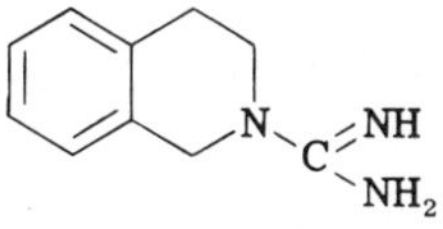

$$RNHC(=NH)NH_2 \xrightarrow[KOH]{\Delta} RNH_2$$

Debrisoquine

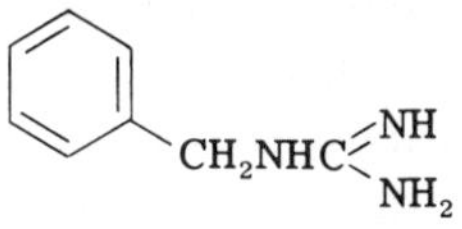

Bethanidine

Cyclohexane was added to the hydrolysis mixture in order to protect the amine by removing it from the strongly alkaline aqueous phase. Hydrolysis of guanethidine and debrisoquine was complete after 4 h, while more than 12 h was needed for guanoxan and bethanidine. Consequently, debrisoquine was chosen as the internal standard for guanethidine.

Biological fluids containing the guanidine drugs were first washed with organic solvent to remove interfering amines: plasma was washed with diethyl ether at pH 7.4, and urine was washed with toluene at pH 10. At these relatively low pH values (for guanidines) the extraction of any of the drugs was negligible and the organic phase could be discarded. The pH of the biological fluid was then increased to over 13.5 and extracted with either chloroform (guanethidine and debrisoquine) or ethylene dichloride (guanoxan and bethanidine). The drugs were back-extracted into 0.1 *M* HCl, which was then made strongly basic by addition of 40% KOH. Finally cyclohexane was added, and the mixture was heated in a screw-capped vial. When hydrolysis was complete, the cyclohexane phase was removed and, after the addition of formic acid, evaporated to dryness. The formic acid was added to prevent any loss of the volatile amines during evaporation. The residue was redissolved in ethyl acetate, and derivatization with TFAA was carried out. Solvent and excess reagent were removed by evaporation, and the residue was dissolved in more ethyl acetate for chromatography. Three stationary phases were used: OV-101, OV-17, and Poly I-110, all 3%

loadings. The column temperature was 130–150°C, depending on the drug being analyzed. When debrisoquine was used as the internal standard, both it and the guanethidine TFA derivatives had retention times of less than 3 min on the Poly I-110 column. The OV-101 phase was used for the more polar derivative of guanoxan; FID or multiple ion detection was used. Levels of guanethidine above 100 ng/ml could be easily analyzed by FID, while multiple ion detection was used for levels below this. For guanethidine an absolute recovery of 48% was obtained from plasma over the range 2 to 60 ng/ml. This paper gives a good example of carefully optimized conditions of extraction and derivatization and illustrates how a difficult problem can be overcome by some bold thinking.

Erdtmansky and Goehl (1975) published a method for guanethidine, debrisoquine and related compounds that combined extraction and derivatization. Plasma or urine, buffered with molar sodium hydrogen carbonate, was heated with benzene and hexafluoroacetylacetone in a test tube fitted with an air condenser. The tube was maintained at 100°C for 2 h, then 3 *M* sodium hydroxide was added, and the mixture was shaken and then centrifuged. A portion of the benzene phase was injected onto a column of 3% OV-17 on Gas Chrom Q (100–120 mesh) at 160°C. The following derivative was formed by the reaction with hexafluoroacetylacetone:

$$RNHC(=NH)NH_2 + CF_3C(=O)CH_2C(=O)CF_3 \longrightarrow \text{RNH-pyrimidine}(4,6\text{-}(CF_3)_2)$$

The authors claimed that their method was capable of measuring nanogram levels of debrisoquine and related drugs (25 ng/ml) without needing the expensive and complicated GC–MS combination.

Another group claiming an even lower limit of detectability with electron-capture detection of guanethidine adopted the hydrolysis procedure of Hengstmann *et al.* and then derivatized with heptafluorobutyric anhydride. The method was rather complex: a portion of the final extract containing the derivative was injected onto a "clean-up" column of 1% DEGS on Chromosorb W-HP (100–120 mesh) and programed from 80 to 190°C at 8°C/min. At the "retention window" for the derivatized guanethidine (and also the standard), a rerouting of the effluent was carried out to an analytical column of 5% OV-225 plus 0.5% Gas Quat-L on Supelcoport (80–100 mesh) at 175°C (Pellizzari and Seltzman, 1979). The authors claim an overall recovery of about 70% with linearity down to 700 pg/ml plasma.

An unusual derivative is described in a method for clonidine measurement, using a capillary column and an EC detector (Edlund, 1980). Clonidine was extracted from alkaline plasma into a mixture of cyclohexane–butanol (9 : 1). The drug was then back-extracted into sulfuric acid, which was made alkaline, and extraction into more cyclohexane–butanol was carried out. The organic phase was then evaporated, and the drug was derivatized with 1% pentafluorobenzyl bromide in acetone with 5–25 mg potassium carbonate present. The mixture was refluxed in a heating block for 45 min. The reaction was as follows:

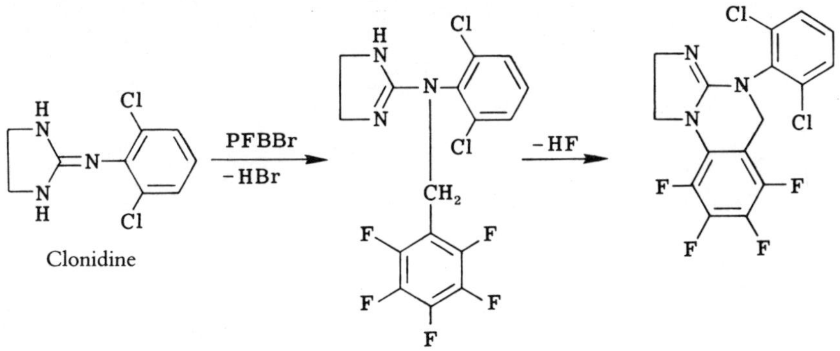

PFBBr
−HBr
−HF
Clonidine

The structure of the derivative was confirmed by mass spectrometry. Gas chromatography was carried out on a platinum–iridium capillary column coated with either OV-17 or SP-2340. With such a method it was necessary to choose as internal standard a compound that closely resembles the drug. In this case 2-[(2,4-dichlorophenyl)amino]-2-imidazole was an excellent choice (see page 121). At the 100 pg/ml level recoveries were in the range 60–80% with a coefficient of variation of 5%.

D. Nifedipine

Nifedipine is one of a relatively new group of drugs that are dihydropyridine derivatives with vasodilating and calcium-antagonist properties. Consequently, such drugs are being used in the treatment of hypertension. Some are characterized by their sensitivity to light, and most workers take precautions to carry out drug analysis using brown-glass test tubes or working under subdued sodium light during the early stages of the analysis particularly in the case of nifedipine. A simple procedure was published by Jakobsen *et al.* (1979) in which plasma was buffered at pH 9, extracted with toluene, and a portion of the centrifuged organic phase injected directly onto a 2% OV-17 column operated at 240°C. The nitro group is strongly electron capturing. The authors claimed that nifedipine was stable on the column, and supported this with GC–MS evidence. However, Kondo *et al.* (1980) reported that, using this method, partial oxidation took place, resulting in two peaks. Accordingly, they went on to develop a method based on a GC–MS method of Higuchi and Shiobara (1978), which involves treating plasma directly with 0.1 *M* HCl and 1% sodium nitrite to convert nifedipine to the corresponding pyridine derivative:

NO_2 $\xrightarrow{HNO_2}$ NO_2

H_3CO_2C CO_2CH_3 H_3CO_2C CO_2CH_3

H_3C N H CH_3 H_3C N CH_3

Nifedipine

The plasma was then extracted with benzene, which was removed and evaporated to dryness, and the residue was redissolved in ethyl acetate for gas chromatography. The column was packed with 3% OV-1 on Chromosorb W (AW-DMCS) (80–100 mesh) and operated at 230°C. Detection down to 5 ng/ml plasma was possible using an electron-capture detector. A very similar method was developed by Rämsch and Sommer (1983) who used gaseous nitrogen dioxide to oxidize the drug to the corresponding pyridine. They claimed a recovery of 102 ± 11% over the range of 5 to 200 ng/ml with a limit of detection of 2 ng/ml. Once oxidation to the pyridine derivative has been carried out, the compound is no longer sensitive to light.

It has recently been demonstrated, however, that the pyridine derivative is actually present in some human plasma specimens as the result of *in vivo* metabolism. Reported nifedipine concentrations, using these methods, which result in this derivative, will therefore be higher than the "true" nifedipine concentration. There has been a move back to measuring nifedipine as the underivatized compound (Dokladalova *et al.*, 1982; Hamann and McAllister, 1983; Lesko *et al.*, 1983). The Jakobsen method (1979) has been found by many to give rather variable results and, especially when using packed columns, some conversion to the pyridine takes place on column. Careful silanizing does not seem to eliminate this completely, and the best solution at present seems to be that adopted by Dokladalova and colleagues, namely, the use of a fused-silica capillary column (OV-101) and an all-glass injection system.

VII. Pentylenetetrazol

This central nervous system stimulant has a rather unusual structure and is soluble in both polar and nonpolar organic solvents. The saturated seven-membered ring and four nitrogen atoms with no replaceable hydrogens

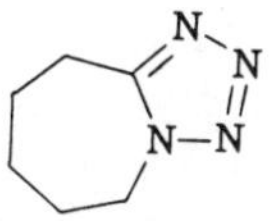

mean that the molecule is very unreactive indeed. It possesses no strongly UV-absorbing chromophores and no native fluorescence, and it is difficult to think of a method available in the 1970s, other than gas chromatography, that would be capable of measuring low concentrations in body fluids. Steward and Story (1972) showed that the drug could be extracted into diethyl ether under alkaline conditions but chose benzene, to reduce emulsion formation, when extracting the drug from blood. After extraction the organic phase was simply evaporated and redissolved in distilled water. When very low levels were to be measured, the residue was redissolved in 10–20 μl distilled water containing procaine hydrochloride. The procaine acted as an internal standard to compensate for any variability in injection technique. The residue from the benzene evaporation could be made up in water because FID was used to quantify the drug. Because of the unusual chemical structure of pentylenetetrazol, both polar and nonpolar stationary phases could be used, but none was entirely satisfactory, and at low concentrations some tailing was encountered. However, 5% PEG-20000 (Carbowax 20 M) was selected as being the most suitable. Overall recoveries were high (98% from urine and over 80% from blood and plasma), and the limit of detection was 25 ng/ml.

VIII. Terodiline

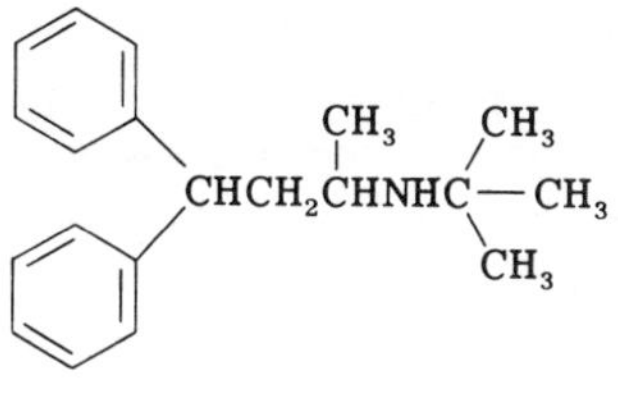

Terodiline is a vasodilator with a secondary amino group in the side chain. Normally, secondary amines are simple to derivatize, but terodiline presents some difficulty because of the steric configuration, the secondary amino group being shielded by two very bulky groups. Conversion to the heptafluorobutyramide would not take place unless a catalyst, such as trimethylamine, was present; even then, its concentration had to be high, 0.15 *M* (Hartvig *et al.*, 1974). The use of catalysts in acylation has been discussed in Chapter 2, and it may be that a 4-dialkylaminopyridine might have been effective at lower concentrations. The derivatization reaction was extremely solvent dependent, the yield being 0% in ethyl acetate, 25% in benzene, and 90% in heptane. The authors used water-saturated solvents and suspected that the low yields might be related to the amount of water held by each solvent: ethyl acetate being the most polar, followed by benzene. To test this, they used three times as much heptane, i.e., increasing the absolute amount of water threefold. The yield of derivative dropped dramatically. The limit of detection of terodiline was 3 ng/ml plasma with a mean recovery of 99% at the 60 ng/ml level.

This method is an excellent illustration of how to control carefully extraction and derivatization conditions to obtain the best results. In such a method it is absolutely essential that an internal standard is chosen that closely mimics the behavior of the drug being measured. The authors used *N-tert*-butyl-1-methyl-3,3-diphenylbutylamine, which has the same steric configuration around the amino group.

IX. Medroxyprogesterone Acetate

Medroxyprogesterone acetate is used to control fertility and was measured by Kaiser *et al.* (1974) by converting it to the 3-enol heptafluorobutyrate ester followed by electron-capture detection. The authors discussed

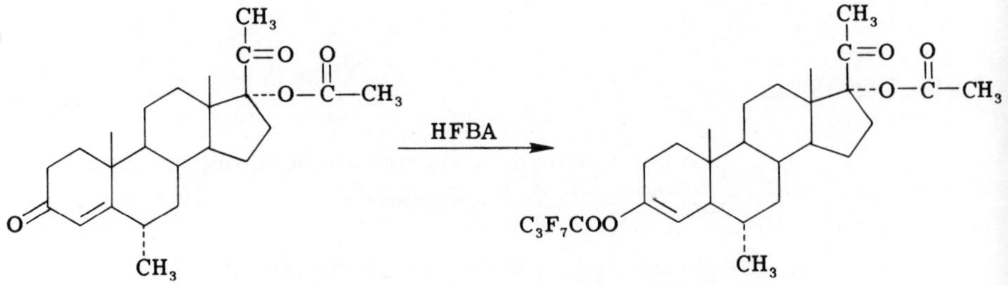

the choice of extracting solvent: the drug was almost quantitatively extracted by cyclohexane, ether, and methyl ethyl ketone. Not surprisingly, they chose cyclohexane because less endogenous material was co-extracted by this nonpolar solvent. 3β-Hydroxy-5-cholenic acid methyl ester was chosen as the internal standard, and this was added after the drug had been extracted. This was not regarded as a serious disadvantage since recoveries were relatively constant over the range measured: 91% at the 5-ng and 103% at the 100-ng level.

A difficulty encountered during the development of the method was the appearance of a second peak in addition to the 3-enol heptafluorobutyrate. Mass spectrometry showed that this peak was produced by a compound with a molecular ion and fragmentation pattern identical to that of the 3-enol derivative, and it was concluded that it was an isomer formed by the migration of the double bond. The amount of this isomer was limited to 5% of the total formed by choosing a derivatization time of 60 min.

X. Phenothiazines

An unusual method of derivatization is that reported by Noonan *et al.* (1972) for the gas chromatography of some phenothiazines using electron-capture detection. Low concentrations of promazine, acepromazine, and triflupromazine were extracted from blood into dichloromethane. The dichloromethane phase was separated, evaporated to dryness, and the residue was redissolved in cyclohexane. The derivatization reaction—bromination—was carried out, free halogen was destroyed by vigorous mixing with 1 *M* NaOH, then the cyclohexane phase was separated by centrifugation, and a portion was chromatographed on a 3% OV-1 column at 220°C for the triflupromazine derivative and 250°C for the other two.

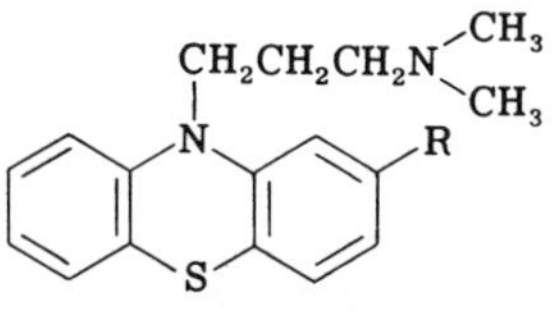

$CH_2CH_2CH_2N(CH_3)_2$ | N, S, R

R	
H	Promazine
$COCH_3$	Acepromazine
CF_3	Triflupromazine

The paper is interesting because it describes the wide range of reaction conditions that were tried. Bromination was performed in a wide range of solvents from the nonpolar cyclohexane through ethyl acetate to acetic acid. A variety of catalysts such as iodine, ferric chloride, and mineral acids were also tried. The best reagent was found to be 1% bromine in iodine-saturated cyclohexane, with pyridinium perbromide in dichloromethane not far behind. The bromination of the phenothiazines was found to be very dependent on the substrate concentration, and at concentrations of 500 ng/ml or less, no reaction took place. Concentration of the substrate by evaporation produced a spectacular increase in derivative yield.

XI. Phanquone

The antiamoebic drug phanquone poses a problem in that it cannot be extracted readily from biological material. In addition, the need for a very sensitive method to study plasma concentrations after a single 50-mg oral dose strongly indicated derivatization. The problem was solved by buffering plasma at pH 3, adding to it 20% w/v aqueous methoxyamine hydrochloride, and heating the solution to 70°C for 2 h. (Degen *et al.*, 1976b). Under these conditions the following reaction took place:

Phanquone $\xrightarrow{CH_3ONH_2 \cdot HCl}$ (dimethoxime: $=NOCH_3$, $=NOCH_3$) $\xleftarrow{CH_3ONH_2 \cdot HCl}$ Metabolite (OH, OH)

The hydroquinone metabolite of phanquone was, unfortunately, also converted to the same derivative, so the separate measurement of both is not possible by this method. However, the metabolite shows the same pharmacological activity as the parent compound. The internal standard, 10-methyl-4,7-phenanthroline-5,6-dione, was added to the plasma before the methoxyamine hydrochloride.

After the reaction had gone to completion, the plasma was made alkaline, and the derivatives were extracted into toluene and then back-extracted into dilute sulfuric acid. The acid phase was separated, made alkaline, and the solution was extracted with fresh toluene. The toluene phase was then removed following centrifugation, evaporated, and the residue was redissolved in a little toluene for chromatography on a 3% JXR column operated at 210°C. Using an electron-capture detector, concentrations of about 15 ng/ml in plasma or urine could be detected. This paper provides a good example of the optimization of the conditions of extraction and derivatization. The reaction of phanquone with methoxyamine hydrochloride was found to be very dependent on pH: at pH 2–5, 85% of the product was the dimethoxime, while at pH 7 the monomethoxime predominated. The derivatives were soluble in a range of common solvents including chloroform and ethyl acetate, but the authors chose toluene, accepting a lower recovery in return for less interference from endogenous biological material.

XII. Itanoxone

The method developed for the estimation of itanoxone, used to treat metabolic disorders, is a good example of the application of a complex procedure to a very difficult problem that might be better solved by a different technique (Cousse *et al.*, 1981). The problem with itanoxone is that it contains a methylene group adjacent to a carboxylic acid group:

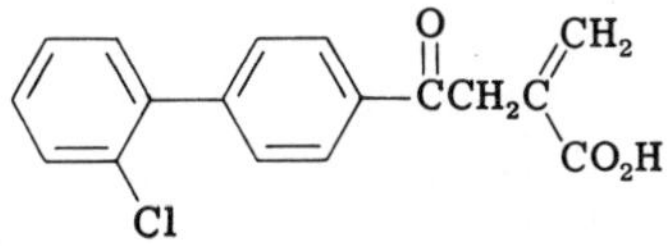

O CH$_2$
CCH$_2$C
CO$_2$H
Cl

The extraction of the drug is simple: plasma buffered at pH 2.2 is extracted with ethyl acetate (for 1–2 h), and the mixture is separated by centrifugation and evaporated to dryness. The residue is redissolved in dioxane, a little acetic acid in dioxane and 2 mg of palladium on charcoal added, and the offending double bond is reduced by passing hydrogen into the solution for 1 h at 0.25 bar. The solution is then centrifuged, the residue extracted with more dioxane, and the fractions are combined and evaporated to dryness. The residue is then suspended in ethereal diazomethane, and esterification is allowed to proceed for 1 h at room temperature. The solvent is then evaporated, the residue dissolved in hexane, and a portion is chromatographed on a column of 3% OV-225 on Chromosorb W AW (80–100 mesh) at 230°C. The 4-chlorophenyl analog was used as the internal standard, and quantitative recovery was claimed. The minimum concentration detected in plasma was 100 ng/ml, presumably owing to the poor electron-capturing properties of the monochlorophenyl group.

Gas chromatography of the unchanged drug led to isomerization on the column:

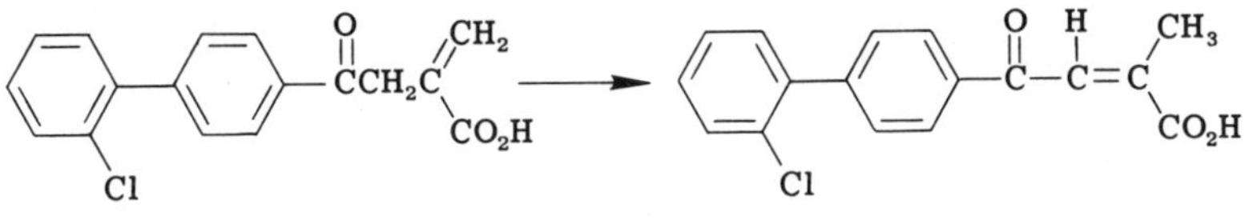

O CH$_2$ O H CH$_3$
CCH$_2$C ⟶ C—C=C
CO$_2$H CO$_2$H
Cl Cl

The methylenic double bond was reduced before esterification to prevent the possible side reaction with diazomethane:

The authors conclude by noting that a simple, rapid HPLC method, which does not require derivatization, has been developed, but they claim that the sensitivity is not as good as that of the GLC method.

XIII. Nonsteroidal Anti-Inflammatory Drugs

Many nonsteroidal anti-inflammatory drugs can be considered as being derived from acetic or propionic acid and, in general, present little difficulty for the analyst. Almost all are very lipid soluble, so they present no problems of extraction, and depending on the concentrations to be measured, they can be chromatographed with or without derivatization. A representative selection is presented here.

A. No Derivatization

Chromatography without derivatization can be carried out when relatively high concentrations are anticipated. Hoffman (1977) measured ibuprofen in serum, using 3-methyl-3-phenylbutyric acid as the internal standard. The serum proteins were first precipitated by adding 10% perchloric acid, and the drug and internal standard were extracted into carbon tetrachloride. The organic phase was separated by centrifugation, and an aliquot was injected directly onto a 5% FFAP column operated at 220°C with flame ionization detection. Under the conditions of the method the retention times of the internal standard and ibuprofen were 2.9 and 4.3 min, respectively. The method was capable of measuring 0.5 μg ibuprofen/ml serum, and Hoffman claimed that the time needed was less than half that for the derivatization method described in Section XIII,C.

B. Derivatization to the Silyl Ester

Fenoprofen has been measured in plasma to which 2-(4-phenoxyphenyl)-valeric acid was added as the internal standard (Nash *et al.*, 1971). The plasma was acidified with 10% trichloroacetic acid and extracted with hexane, which was removed following centrifugation. The plasma was extracted with more hexane, and the organic phases were combined and back-extracted into 0.1 *M* sodium hydroxide. The pH was adjusted to 3 with acetic acid, and the drug and internal standard were extracted into hexane. The organic phase was separated, evaporated to dryness, and silylation was carried out with 20 μl of 5% HMDS in carbon disulfide. The mixture was sonicated for 30 sec, and a portion was chromatographed directly on 3.8% UCCW-982 on Diatoport S (80–100 mesh) at 175°C. At a concentration of 5 μg/ml the coefficient of variation was 5.5%.

C. Derivatization to the Methyl Ester

Kaiser and Vangiessen (1974) determined ibuprofen in plasma after esterification. Acidified plasma was extracted with benzene, which was separated and evaporated to dryness. Conversion to the methyl ester was carried out very rapidly (about 1 min), using 1,1′-carbonyldiimidazole in chloroform (see page 44). The chloroform also contained naphthalene as the internal standard for gas chromatography. Methanol was added to the reaction mixture to complete the esterification, and then the mixture was made alkaline, shaken, and centrifuged. A portion of the chloroform phase was used directly for chromatography on a 6% DEGS column at 150°C with flame ionization detection. The average recovery was 95%, and 0.5 μg ibuprofen/ml plasma could be detected. The retention times of naphthalene and ibuprofen methyl ester were 3.4 and 6.8 min, respectively.

D. Derivatization to the Pentafluorobenzyl Ester

In the method for ibuprofen, in which the drug was converted to the pentafluorobenzyl ester, Kaiser and Martin (1978) used ibufenac (4-isobutylphenylacetic acid) as the internal standard. Serum was made acidic, and the drug and internal standard were extracted into benzene, which was then removed and evaporated to dryness. The residue was redissolved in a small volume of chloroform, and TLC was carried out. The developed plate was examined at 254 nm, and the ibuprofen and ibufenac zones were scraped off, and the drugs were eluted into methanol. This was evaporated, and the drugs were esterified with pentafluorobenzyl bromide in acetone with a little potassium carbonate for 1 h at 60°C. The mixture was then evaporated, and the derivatives were dissolved in cyclohexane. The cyclohexane was removed, evaporated, and the residue was suspended in a small volume of fresh cyclohexane (0.2 ml) for GLC on a column of 10% w/v 3-cyanopropylsilicone on Gas Chrom Q (100–120 mesh) at 190°C. The retention times of the ibuprofen and ibufenac esters were 9.1 and 13.3 min, respectively. The drug could be estimated at concentrations of 0.1 μg/ml of serum, and the overall recovery was 98%. This method is rather complex, and TLC purification is extremely time consuming and is not to be recommended if large numbers of specimens are to be analyzed.

Some nonsteroidal anti-inflammatory drugs are given in relatively low doses and are extensively metabolized; in such cases derivatization and electron-capture detection are usually necessary. This is the case with the anti-inflammatory drug diclofenac, which is derivatized by esterification leading to ring closure (Geiger *et al.*, 1975b):

CO_2H NH Cl Cl → $CO_2CH_2CF_3$ NH Cl Cl → N O Cl Cl

The carboxylic acid group is converted to the trifluoroethyl ester, and under alkaline conditions the ring closes smoothly to give the indolone. The ring closure is carried out simply by adding hexane and enough saturated potassium hydrogen carbonate to make the reaction mixture alkaline. The mixture is shaken for only 15 sec, and the indolone is extracted into the hexane phase. A portion of this is evaporated, redissolved in a little hexane, and chromatographed on a column of 3% JXR on Gas Chrom Q operated at 205°C. Concentrations of 2 ng/ml of sample could be detected. A methoxy analog of diclofenac was used as the internal standard, and this compound also underwent conversion to the corresponding indolone.

An FID method was published by Brombacher *et al.* (1977), but this was not capable of measuring concentrations below 100 ng/ml plasma or serum. This is unsatisfactory for detailed pharmacokinetic studies, which require blood levels to be monitored over 12 h. Interestingly, the authors treat diclofenac with acetyl chloride in benzene. No claim is made as to the structure of the derivative, but it is almost certainly converted to the same indolone as in the Geiger method. Conversion of diclofenac to the methyl ester has also been reported, and this possesses a greater electron-capturing response, but care must be taken with the reagents used since conversion to the indolone is possible if the pH is not carefully controlled (Ikeda *et al.*, 1980; Jack and Willis, 1981).

Degen and Schneider (1983) have reported the interesting simultaneous derivatization of carboxyl and phenolic groups in the new analgesic and antiphlogistic compound CGP 6258. The heptafluorobutyric anhydride,

CO_2H $C_3F_7O_2C$ $CO_2CH_2CF_2CF_3$

HFBA

$CF_3CF_2CH_2OH$
70°C, 1 h

which is the acylating agent, is also the catalyst for the esterification with 2,2,3,3,3-pentafluoro-1-propanol. The selectivity of the method is such that no purification step is necessary after derivatization. The reaction mixture is evaporated to dryness, and the residue is dissolved in *n*-heptane for chromatography on a 3% OV-225 column (Supelcoport, 80–100 mesh) operated at 220°C.

XIV. Antidepressants

The determination of the most commonly used antidepressants presents little problem when using gas chromatography. The drugs are generally very lipid soluble and easily extracted into nonpolar organic solvents. The most difficult problem is deciding which groups of patients to monitor and how to interpret the results of this monitoring. In general, monitoring serum and plasma levels seem most useful in those patients who are taking their therapy as directed but fail to respond. Titrating the dose to the individual patient is more difficult with antidepressants than with many other classes of drug because weeks are often needed for the full therapeutic effect to become apparent. There are also the well-known difficulties in accurately and reproducibly assessing the effect of such treatment.

The decision whether or not to derivatize depends on the nature of the drug, and since many antidepressants are tertiary amines, it is often possible to chromatograph them directly. However, if low doses are being given and plasma concentrations have to be monitored accurately over many hours, it may be desirable to convert the drug to a strongly electron-capturing derivative. Many of the methods published attempt to measure metabolites as well as unchanged drug, and since N-demethylation is a common pathway, a derivatization step is usually included. The measurement of metabolites is discussed in Chapter 7. Only a selection of the many available methods are presented here, and if a more detailed treatment is desired, the reviews by Gupta and Molnar (1980) and Riess *et al.* (1979) are worth studying. The latter review a number of aspects of antidepressant monitoring and includes a survey of attempts to correlate plasma levels with clinical effect. It is important to remember that since many antidepressants are tertiary amines, adsorption to glass and other material is a potential problem. To reduce this hazard, glassware should be rigorously cleaned and silanized, and this process should always be carried out under standardized conditions.

A. Amitriptyline

Amitriptyline has been measured in serum, using the pyrrolidine analog as the internal standard with N-FID. In the method of Jørgensen (1975), serum is buffered at pH 9, and the drug (together with its principal metabolite nortriptyline) and internal standard are extracted with hexane. The organic phase is removed, evaporated, and the residue is dissolved in 0.5 ml of hexane containing 0.1% triethylamine. A 5-μl volume of a 2% solution of acetic anhydride in hexane is added, and the solution is evaporated to dryness. The acetic anhydride is added to convert the nortriptyline to the acetamide derivative since amitriptyline itself, a tertiary amine, is not derivatized in this reaction. The residue is redissolved in a small volume of triethylamine–hexane, and a portion is chromatographed on a column of 1% OV-17 on Chromosorb CQ (100–120 mesh) at 234°C. Under the chromatographic conditions used, the retention times of amitriptyline and internal standard were 1.5 and 3.8 min, respectively. The retention times of nortriptyline and its acetamide derivative were also given. The lower limit of detection for amitriptyline was 5 ng/ml, and no interference was encountered from metabolites or a number of commonly used benzodiazepines. The extraction pH of 9 used in this method is lower than that used in several other methods for measuring this antidepressant. Many methods for amitriptyline also provide a separate measurement of nortriptyline, and it has been shown that combined amitriptyline and metabolite concentrations exhibit a better correlation with the management of depression than the concentration of either compound alone (Braithwaite *et al.*, 1972).

B. Amitriptyline with Derivatization and EC Detection

In the method for amitriptyline involving derivatization and EC detection, developed by Wallace *et al.* (1975), the drug is extracted from alkaline plasma or serum with hexane. The organic phase is shaken with a ceric sulfate–sulfuric acid solution, which removes the drug from the hexane, and when the ceric sulfate-containing phase is heated to 95°C for 10 min, amitriptyline is converted to anthraquinone:

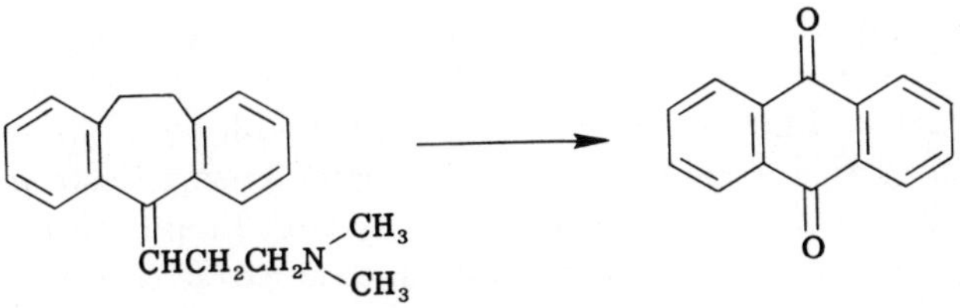

O

CH_3

$CHCH_2CH_2N$

CH_3

O

On cooling, the mixture is extracted with 0.1 ml hexane containing the internal standard ethylanthraquinone, and GC is carried out on 3% OV-17 on Gas Chrom Q (100–120 mesh) at 210°C. This method is capable of measuring 2 ng/ml using 0.5 ml of specimen. The active metabolite, nortriptyline, is also converted to anthraquinone and cannot be measured separately; however, as we have seen in the last paragraph, this is not necessarily a disadvantage.

C. Maprotiline

Maprotiline is unlike many antidepressants in that it has a tetracyclic rather than a tricyclic structure. It is measured in biological fluids after derivatization because of the relatively low levels present and because the underivatized drug, which contains a secondary amino group, tails on the column at low concentrations. Gupta *et al.* (1977) produced a method for measuring serum maprotiline using N-FID. Desmethyldoxepin was added to serum as the internal standard, and the serum was made acid and washed with pentane. The pentane layer was removed and discarded, and the serum was made alkaline with sodium hydroxide. Extraction was carried out with more pentane, which was then removed and transferred to silanized tubes. To these tubes were also added 10 μl acetic anhydride and 5 μl pyridine, and the pentane was evaporated on a water bath at 40–45°C. The "dry" tubes were then placed in an oven at 100°C to remove any remaining traces of acetic anhydride and pyridine. The residue in each tube was redissolved in 20 μl of methanol for chromatography on a 3% column of HI-EFF-8BP on Gas Chrom Q (80–100 mesh) at 240°C. Under these conditions, the retention times of the acetyl derivatives of desmethyldoxepin and maprotiline were 9.5 and 12.3 min, respectively. Concentrations of 10 ng/ml, starting with 3 ml serum, could be measured. The coefficient of variation at 100 ng/ml was 11.5% between batches.

The authors compared other stationary phases such as SE-30, OV-17, XE-60, and OV-225, all 3% loadings. The relatively nonpolar SE-30 and OV-17 did not separate well the desmethylmaprotiline derivative from that of maprotiline, and this was obviously unsatisfactory. The polar XE-60 and OV-225 were satisfactory in this respect, but because they contained CN groups a higher background bleed was observed with the N-FID.

Geiger *et al.* (1975a) produced a sensitive method using electron-capture detection with nortriptyline as the internal standard. Blood was buffered at pH 10, and the drug and internal standard were extracted into heptane–isopropanol (99:1). The amines were back-extracted into dilute sulfuric acid, which was then basified, and the solution was extracted with hexane. The hexane was removed and evaporated not to dryness but until the volume was reduced to ~50 μl. To this was added 100 μl of freshly prepared acetonitrile–heptane–heptafluorobutyric anhydride (1:93:6). Derivatization was allowed to proceed at 70–75°C for 90 min. The reaction mixture

CH_2–CH_2–CH_2–N(CH_3)(H) —HFBA→ CH_2–CH_2–CH_2–N(CH_3)(COC_3H_7)

was then evaporated, and the residue was redissolved in about 1 ml hexane and shaken with 0.5 ml 1 *M* NaOH for 20 sec. The hexane phase was removed, evaporated to dryness, and redissolved in fresh hexane for chromatography on a column of 3% JXR on Gas Chrom Q at 230°C. The method was claimed to be able to detect less than 10 ng per biological sample, and the retention times for the heptafluorobutyryl derivatives of nortriptyline and maprotiline were 3 and 4.5 min, respectively. A comparison was made between this method and the double radioisotope derivative assay described by Riess (1974): both methods gave almost identical results.

XV. Quality Control

No treatment of the measurement of drugs in biological fluids would be complete without a discussion of quality control. No matter how good the analytical method and how sophisticated the equipment, the results produced will be valueless unless some means of assessing their precision and accuracy is used. In the field of drug analysis it was not an uncommon belief that one instinctively "knew" that results were meaningful, based on past experience: quality control schemes were "elaborate" and "time wasting." This attitude is, thankfully, becoming rarer. Good quality control is essential, and it is important not only that a method should function well, but that it should be seen to do so.

Quality control is simply an objective means of assessing the end product of some particular process: the product may be an automobile, a watch, or a razor blade. The concentration of potassium in a patient's serum or the amount of propranolol per ml plasma is the product of an analytical procedure. The need for good quality control was recognized very early and soon became widespread in manufacturing industries. However, clinical chemists, generally recognized as the front runners in laboratory quality control, got started rather late in the game (1950s), as judged by the appearance of publications dealing with quality control as a topic in its own right. Those workers involved in drug analysis in body fluids were even slower off the mark, and there may be still some who have never left the starting blocks.

Control of the quality of the results of drug analysis is relatively simple since one can start with blank blood, plasma, serum, etc. and accurately add known amounts of drug to it. The task facing clinical chemists is more difficult, since almost all their measurements are concerned with compounds already present in biological fluids. There is no simple way in which they can start with blank serum if they wish to measure calcium for example. Because their problems are more complex, clinical chemists have devoted more time to their solution, and it is well worth reading the excellent introduction to the subject by Whitehead (1977) or the text by Ottaviano and DiSalvo (1977). The application of quality control to toxicology has been examined in a work edited by Paget (1977), and a very good review of quality control in the measurement of drugs has been published by Griffiths *et al.* (1980).

To illustrate some aspects of quality control, let us consider as a concrete example the analysis of the β-adrenoceptor blocker metoprolol in patient plasma. Each time a batch of patient plasma specimens has to be analyzed, a series of standards will be incorporated in the run to allow a standard curve to be constructed. This should be done with every batch and must be done when complex extraction procedures with derivatization are used. The calibration curves produced with each batch should be kept as a record of how the method was working at any particular time. In addition to this, it is advisable to incorporate a number of additional specimens (quality control, or QC, specimens) into each batch, their exact number depending on the size of the batch to be analyzed. As a rough guide it is a good idea to have a couple of QC specimens for every 10 patient specimens. These are simply blank plasma specimens containing known added amounts of metoprolol, half usually containing a low and the other half a high concentration. The calibration curve is constructed from the results obtained from the standards (note: the QC specimens are not used in the construction of the calibration curve), and the concentration of metoprolol in the patient specimens and in the QC specimens are read from this curve. The results of the metoprolol analysis of the patient specimens can be accepted if the calibration curve is satisfactory, i.e., if it is linear over the relevant range of metoprolol concentration and if the results for the QC specimens are within acceptable limits, usually ±10% of the known values. Occasional anomalous results for standards and QC specimens can be rejected if there is a known cause, e.g., a tube was dropped and most of the contents lost. However, if more than one standard is rejected, it is safer to repeat the analysis of the batch.

Using this procedure, each batch of results will include a low and a high QC, and these should be kept for reference at a later date and quoted every time the results of that particular batch are used. The high QC should be chosen such that it is close to the maximum plasma concentration likely to be encountered, while the low QC should be about one-tenth of this concentration. More QC specimens can, of course, be incorporated if desired, but a balance must be struck between using too few and swamping a handful of patient specimens with a multitude of standards and QCs.

This represents an adequate quality control procedure. If the drug being studied is relatively stable, about 20 ml of plasma can be spiked with drug and 1-ml portions transferred to vials and deep-frozen. At regular intervals, say weekly, portions can be analyzed to give what is called a repeat analysis check or a sample from one batch reanalyzed with the next.

Where possible, and especially when a new method is being set up, interlaboratory comparisons should be carried out. These comparisons can be done either with the company manufacturing the drug or another laboratory experienced in its analysis. If different techniques are used, say GLC in one laboratory and HPLC in another, the results may not be exactly similar, however, a good correlation should be sought. Gas chromatographic methods are often taken as reference methods since they are less susceptible to interference from other drugs and endogenous components.

If relatively new drugs are being measured, then the opportunities for interlaboratory comparisons are limited, but for the more frequently used established drugs a number of QC schemes may be in operation. For example, there is the AED scheme for antiepileptic drugs, which includes phenytoin, phenobarbital, primidone, carbamazepine, valproic acid, ethosuximide, and clonazepam; the RESP scheme for theophylline; and the PSYCH scheme for imipramine, desmethylimipramine, amitriptyline, and nortriptyline. These external QC schemes grew out of the St. Bartholomew's Hospital Quality Control Scheme for Antiepileptic Drugs in the United Kingdom, and the schemes are now referred to under the general name of Bartscontrol. On the first day of each month two serum specimens are sent

```
SAMPLE: PPP 879
DRUG: PHENYTOIN
METHOD: GLC  -  NO DERIVATIVE

DRUG     NO
LEVEL

           ********************************
  0.00    0*                              *
  4.00    0*                              *
  8.00    0*                              *
 12.00    0*                              *
 16.00    0*                              *
 20.00    0*                              *
 24.00    1*X                             *
 28.00    1*X                             *
 32.00    1*X                             *
 36.00    0*                              *
 40.00    1*X                             *
 44.00    1*X                             *
 48.00    3*XX                            *
 52.00    5*XXXX                          *
 56.00    9*XXXXXX                        *
 60.00   15*XXXXXXXXXXX                   *
 64.00   31*XXXXXXXXXXXXXXXXXXXXXX <<<<<<*
 68.00   38*XXXXXXXXXXXXXXXXXXXXXXXXXXXX  *
 72.00   27*XXXXXXXXXXXXXXXXXXXX          *
 76.00   12*XXXXXXXXX                     *
 80.00    7*XXXXX                         *
 84.00    5*XXXX                          *
 88.00    3*XX                            *
 92.00    1*X                             *
 96.00    0*                              *
100.0     0*                              *
104.0     1*X                             *
108.0     0*                              *
112.0     1*X                             *
116.0     0*                              *
           ********************************

   NUMBER ACCEPTED                158
   NUMBER REJECTED                  7
   MEAN OF ACCEPTED RESULTS     69.32
   STANDARD DEVIATION            8.71
   COEFFICIENT OF VARIATION     12.57%
   MINIMUM                      41.00
   MAXIMUM                      93.90
   MEDIAN                       69.47
   SPIKED VALUE                 71.60
   YOUR RESULT (<<<<)           67.90
   S.D. S FROM MEAN              0.16
  ********************************
  *  DRUG PERFORMANCE INDEX:  2   *
  ********************************
```

FIG. 6.7. Histogram indicating a laboratory's performance. From Griffiths *et al.* (1980).

out to all the participating laboratories, one containing a low and the other a high concentration of the relevant drugs.

The specimens are prepared from a pool of serum that has been shown to be negative for hepatitis B surface antigen (HB_sAg) and the relevant drugs have been weighed in. The specimens were originally dispatched in the fluid state with sodium azide added as a preservative, but problems with leakage and the potential interference of sodium azide with some methods led to the change to freeze-dried specimens. The specimens are reconstituted in the individual laboratories, analyzed, and the results are entered on special sheets, which are returned to Bartscontrol. Results are collated under specific methods, e.g., phenytoin results will be collected in four separate categories: GLC—no derivative, GLC—derivative, HPLC, and EMIT.

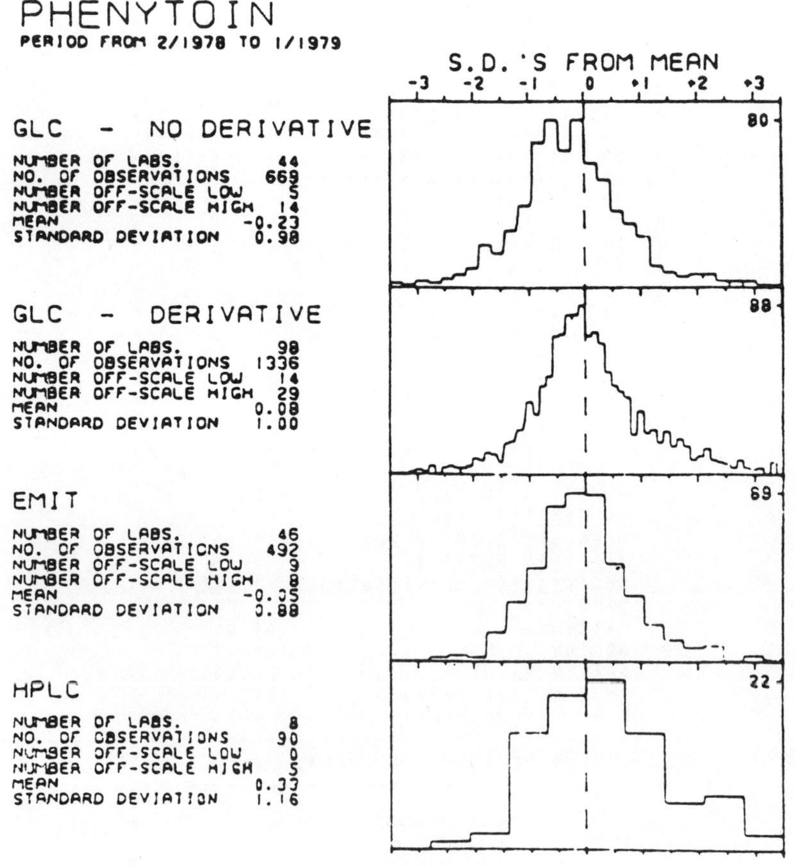

FIG. 6.8. Yearly printout for GLC detection of phenytoin. From Griffiths *et al.* (1980).

Each laboratory receives a computer printout in histogram form indicating the laboratory's performance in relation to others using a similar method. An example of such a histogram is shown in Fig. 6.7. The names of the other laboratories participating in the scheme are not revealed to individual users.

This computer printout lists the mean, coefficient of variation, and other relevant details as well as the individual laboratory's standard deviation from the mean for this particular month. Results that are more than 3.5 standard deviations from the mean are rejected. A drug performance index is also given, and this helps introduce a competitive spirit into the scheme: an index of 1 indicates the laboratory is in the top 10% for that particular month, while an index of 10 relegates the laboratory to the lowest 10%. It should be remembered that these schemes relate the performance of an individual laboratory to the other laboratories participating. Yearly printouts for particular methods are also issued and one of these is shown in Fig. 6.8.

Experience with such schemes has shown that the performance of many laboratories improves markedly, and unsatisfactory methods are rapidly identified. Although initial experience in such a scheme can be nerve-racking, most analysts would probably agree that to be put on one's metal once a month results in an overall improved performance and increased confidence.

References

Borg, K. O., Gabrielsson, M., and Jönsson, T. E. (1974). *Acta Pharm. Suecica* **11,** 313.

Braithwaite, R. A., Goulding, R., Theano, G., Bailey, J., and Coppen, A. (1972). *Lancet* **1,** 1297.

Brazier, J. L., and Delaye, D. (1981). *J. Chromatogr.* **224,** 439.

Brombacher, P. J., Cremers, H. M. H. G., Verheesen, P. E., and Quanjel-Schreurs, R. A. M. (1977). *Arzneim. Forsch.* **27,** 1597.

Caddy, B., and Stead, A. H. (1977). *Analyst,* **102,** 42.

Cook, N. J., and Jowett, D. A. (1983). *J. Chromatogr.* **272,** 181.

Cooper, T. B., Robinson, D. S., and Nies, A. (1978). *Commun. Psychopharmacol.* **2,** 505.

Cousse, H., Ribet, J-P., and Vezin, J-C. (1981). *J. Pharm. Sci.* **70,** 860.

De Bruyne, D., Kinsun, H., Moulin, M. A., and Bigot, M. C. (1979). *J. Pharm. Sci.* **68,** 511.

Degen, P. H. (1979). *J. Chromatogr.* **176,** 375.

Degen, P. H., and Riess, W. (1976). *J. Chromatogr.* **117,** 399.

Degen, P. H., and Schneider, W. (1983). *J. Chromatogr.* **277,** 361.

Degen, P. H., and Schweizer, A. (1977). *J. Chromatogr.* **142,** 549.

Degen, P. H., Schneider, W., Vuillard, P., Geiger, U. P., and Riess, W. (1976a). *J. Chromatogr.* **117,** 407.

Degen, P. H., Brechbühler, S., Schäublin, J., and Riess, W. (1976b). *J. Chromatogr.* **118,** 363.

Degen, P. H., Brechbühler, S., Schneider, W., and Zbinden, P. (1982). *J. Chromatogr.* **233,** 375.

Dekker, W. J., Combs, H. F., and Corby, D. G. (1968). *Toxicol. Appl. Pharmacol.* **13,** 454.
Di Salle, E., Baker, K. M., Bareggi, S. R., Watkins, W. D., Chidsey, C. A., Frigerio, A., and Morselli, P. L. (1973). *J. Chromatogr.* **84,** 347.
Dokladalova, J., Tykal, J. A., Coco, S. J., Durkee, P. E., Quercia, G. T., and Korst, J. J. (1982). *J. Chromatogr.* **231,** 451.
Edlund, P-O. (1980). *J. Chromatogr.* **187,** 161.
Edlund, P-O., and Paalzow, L. K. (1977). *Acta Pharm. Toxicol.* **40,** 145.
Edwards, K. D. G., and McCredie, M. (1967). *Med. J. Austral.* **1,** 534
Erdtmansky, P., and Goehl, T. J. (1975). *Analyt. Chem.* **47,** 750.
Ervik, M. (1969). *Acta Pharm. Suecica* **6,** 393.
Ervik, M. (1975). *Acta Pharmacol. Toxicol.* **36,** Suppl. V, 136.
Ervik, M., and Gustavii, K. (1974). *Anal. Chem.* **46,** 39.
Ervik, M., Kylberg-Hanssen, K., and Lagerström, P. (1980). *J. Chromatogr.* **182,** 341.
Fellenberg, A. J., and Pollard, A. C. (1977). *Clin. Chim. Acta* **81,** 203.
Fleuren, H. L. J., and Van Rossum, J. M. (1978). *J. Chromatogr.* **152,** 41.
Fransson, B., and Schill, G. (1975). *Acta Pharm. Suecica* **12,** 107.
Ganansia, J., Gillet, G., Padovani, P., and Bianchetti, G. (1983). *J. Chromatogr.* **275,** 183.
Geiger, U. P., Rajagopalan, T. G., and Riess, W. (1975a). *J. Chromatogr.* **114,** 167.
Geiger, U. P., Degen, P. H., and Sioufi, A. (1975b). *J. Chromatogr.* **111,** 293.
Griffiths, A., Hebdige, S., Perucca, E., and Richens, A. (1980). *Therap. Drug Monitoring* **2,** 51.
Gudzinowicz, B. J. (1967). "Gas Chromatographic Analysis of Drugs and Pesticides." Dekker, New York.
Guerret, M., Lavene, D., and Kiechel, J. R. (1980). *J. Pharm. Sci.* **69,** 1191.
Gupta, R. N., and Molnar, G. (1980). *Biopharmaceut. Drug Disposit.* **1,** 259.
Gupta, R. N., Molnar, G., and Gupta, M. L. (1977). *Clin. Chem.* **23,** 1849.
Gupta, R. N., Eng, F., and Gupta, M. L. (1979). *Clin. Chem.* **25,** 1303.
Gustavii, K. (1967). *Acta Pharm. Suecica* **4,** 233.
Gyllenhaal, O., and Albinsson, A. (1978). *J. Chromatogr.* **161,** 343.
Gyllenhaal, O., and Vessman, J. (1983). *J. Chromatogr.* **273,** 129.
Hamann, S. R., and McAllister, R. G. (1983). *Clin. Chem.* **29,** 158.
Hartvig, P., Freij, G., and Vessman, J. (1974). *Acta Pharm. Suecica* **11,** 97.
Heipertz, R., Pilz, H., and Eickhoff, K. (1977). *Clin. Chim. Acta* **77,** 307.
Hengstmann, J. H., Falkner, F. C., Watson, J. T., and Oates, J. (1974). *Anal. Chem.* **46,** 34.
Higuchi, S., and Shiobara, Y. (1978). *Biomed. Mass Spectrom.* **5,** 220.
Hoffman, D. J. (1977). *J. Pharm. Sci.* **66,** 749.
Huggett, A., Andrews, P., and Flanagan, R. J. (1981). *J. Chromatogr.* **209,** 67.
Hulshoff, A., and Roseboom, H. (1979). *Clin. Chim. Acta* **93,** 9.
Ikeda, M., Kawase, M., Hiramatsu, M., Hirota, K., and Ohmori, S. (1980). *J. Chromatogr.* **183,** 41.
Jack, D. B. (1973). GLC determination of β-adrenoceptor blocking agents in plasma. Unpublished work.
Jack, D. B. (1978). Unpublished work.
Jack, D. B. (1981). *J. Pharm. Sci.* **70,** Open Forum IV (March).
Jack, D. B., and Laugher, S. J. (1982). Unpublished work.
Jack, D. B., and Riess, W. (1974). *J. Chromatogr.* **88,** 173.
Jack, D. B., and Willis, J. V. (1981). *J. Chromatogr.* **223,** 484.
Jack, D. B., Stenlake, J. B., and Templeton, R. (1972). *Xenobiotica* **2,** 35.
Jack, D. B., Brechbühler, S., Degen, P. H., Zbinden, P., and Riess, W. (1975). *J. Chromatogr.* **115,** 87.

Jakobsen, P., Lederballe Pedersen, O., and Mikkelsen, E. (1979). *J. Chromatogr.* **162,** 81.
Jørgensen, A. (1975). *Acta Pharmacol. Toxicol.* **36,** 79.
Kaiser, D. G., Carlson, R. G., and Kirton, K. T. (1974). *J. Pharm. Sci.* **63,** 420.
Kaiser, D. G., and Martin, R. S. (1978). *J. Pharm. Sci.* **67,** 627.
Kaiser, D. G., and Vangiessen, G. J. (1974). *J. Pharm. Sci.* **63,** 219.
Kates, R. E., and Jones, C. L. (1977). *J. Pharm. Sci.* **66,** 1490.
Kaye, C. M. (1980). *In* "Progress in Drug Metabolism" (J. W. Bridges and L. F. Chasseaud, ed.), Vol. 4, p. 165. Wiley, London.
Kinney, C. D. (1981). *J. Chromatogr.* **225,** 213.
Kondo, S., Kuahiki, A., Yamamoto, K., Akimoto, K., Takahasi, K., Awata, N., and Sugimoto, I. (1980). *Chem. Pharm. Bull.* **28,** 1.
Le Petit, G. (1977). *Pharmazie* **32,** 289.
Le Petit, G. (1980). *Pharmazie* **35,** 696.
Lesko, L. J., Miller, A. K., Yeager, R. L., and Chatterji, D. C. (1983). *J. Chromatogr. Sci.* **21,** 415.
Marzo, A., Quadro, G., and Treffner, E. (1983). *J. Chromatogr.* **272,** 95.
Meola, J. M., and Vanko, M. (1974). *Clin. Chem.* **20,** 184.
Modin, R., and Schill, G. (1975). *Talanta* **22,** 1017.
Nash, J. F., Bopp, R. J., and Rubin, A. (1971). *J. Pharm. Sci.* **60,** 1062.
Noonan, J. S., Blake, J. W., Murdick, P. W., and Ray, R. S. (1972). *Life Sci.* **11,** 363.
Ottaviano, P. J., and Di Salvo, A. F. (1977). "Quality Control in the Clinical Laboratory." University Park Press, Baltimore, Maryland.
Paget, G. E. (ed.). (1977). "Quality Control in Toxicology." MTP Press Ltd., Lancaster, England.
Pellizzari, E. D., and Seltzman, T. P. (1979). *Anal. Biochem.* **96,** 118.
Plavšić, F. (1978). *Clin. Chim. Acta* **88,** 551.
Rämsch, K. D., and Sommer, J. (1983). *Hypertension* **5,** Suppl. II, 18.
Rawlins, M. D., Collste, P., Frisk-Holmberg, M., Lind, M., Östman, J., and Sjöqvist, F. (1974). *Eur. J. Clin. Pharmacol.* **7,** 353.
Riess, W. (1974). *Anal. Chim. Acta* **68,** 363.
Riess, W., Brechbühler, S., and Dubois, J. P. (1979). *In* "Progress in Drug Metabolism," Vol. 3, p. 115. Wiley, London.
Reece, P. A., Stanley, P. E., and Zacest, R. (1978). *J. Pharm. Sci.* **67,** 1150.
Reid, E. (1976). *Analyst* **101,** 1.
Sadee, W., and Beelen, G. C. M. (1980). "Drug Level Monitoring." Wiley, New York.
Scales, B., and Copsey, P. B. (1975). *J. Pharm. Pharmacol.* **27,** 430.
Schill, G., Borg, K. O., Modin, R., and Persson, B. A. (1977). *In* "Progress in Drug Metabolism," Vol. 2, p. 219. Wiley, London.
Serfontein, W. J., and De Villiers, L. S. (1977). *J. Chromatogr.* **130,** 342.
Sioufi, A., Leroux, F., and Sandrenan, N. (1983a). *J. Chromatogr.* **272,** 103.
Sioufi, A., Colussi, D., and Mangoni, P. (1983b). *J. Chromatogr.* **278,** 185.
Stewart, J. T., and Story, J. L. (1972). *J. Pharm. Sci.* **61,** 1651.
Tocco, D. J., de Luna, F. A., and Duncan, A. E. W. (1975). *J. Pharm. Sci.* **64,** 1879.
Trevor, A., Rowland, M., and Leong Way, E. (1972). *In* "Fundamentals of Drug Metabolism and Drug Disposition" (B. N. La Du, H. G. Mandel, and E. Leong Way, eds.), p. 369. Williams & Williams, Baltimore.
Van Boven, M., and Daenens, P. (1979). *J. Forensic Sci.* **24,** 55.
Wallace, J. E., Hamilton, H. E., Goggin, L. K., and Blum, K. (1975). *Anal. Chem.* **47,** 1516.
Walle, T. (1974). *J. Pharm. Sci.* **63,** 1885.

Wan, S. H., Maronde, R. F., and Matin, S. B. (1978). *J. Pharm. Sci.* **67,** 1340.
Whitehead, T. P. (1977). "Quality Control in Clinical Chemistry." Wiley, New York and London.
Wong, K. P., Ruthven, C. R. J., and Sandler, M. (1973). *Clin. Chim. Acta* **47,** 215.
Zak, S. B., Lukas, G., and Gilleran, T. G. (1977). *Drug Met. Disposit.* **5,** 116.
Zak, S. B., Honc, F., and Gilleran, T. G. (1980). *Analyt. Lett.* **13,** 1359.

Chapter 7

Measurement of Metabolites

The many pathways of drug matabolism are classified under two headings: Phase 1 and Phase 2. Phase 1 biotransformations are nonsynthetic reactions such as hydroxylation, dealkylation, reduction, and hydrolysis, while Phase 2 transformations are synthetic and include glucuronic acid conjugation, ethereal sulfate conjugation, methylation, and acetylation. It would be outside the scope of this chapter to give an account of the range of transformations employed by Nature, intruiging though they are, and the interested reader would do well to consult the works by Parke (1968) or Curry (1980).

The principles and techniques applied to measure drug metabolites in biological fluids do not differ in any important respect from those used to determine the parent drugs. In cases where the metabolite does not differ greatly in polarity from the drug, it is often possible to determine both on the same column under the same chromatographic conditions. For example, if the drug is a tertiary amine and is metabolized by dealkylation, the mono- and didesmethyl metabolites will often be separable on a single column and such a case is described on page 184. If, however, a drug containing a side chain undergoes oxidative metabolism, resulting in the corresponding alcohol, the difference in polarity will be such that it is unlikely that both may always be measured under the same conditions. An example of this is given on page 173. Measurement of metabolites is most important when they have a pharmacological effect or when they may produce unwanted and toxic side effects. It is, however, important to remember that normally low levels of a metabolite can increase rapidly with deteriorating renal function.

The products of Phase 2 transformations, particularly glucuronide and sulfate conjugates, are extremely water soluble and consequently cannot be extracted readily by nonpolar organic solvents. Glucuronides can be removed from biological fluids using ion-exchange resins such as XAD-2. Glucuronide or sulfate conjugates can be hydrolyzed by acid and can also be split enzymatically using β-glucuronidase and sulfatase. Resulting aglycones can usually then be extracted using organic solvents and determined by GLC. However, there are objections to splitting conjugates using such methods. For example, acid hydrolysis may split a number of different conjugates, and evidence for a glucuronide or sulfate conjugate may then simply be tentative. The use of strong acid may also destroy the liberated aglycone. Biological fluids may contain different amounts of enzyme inhibitors or alternative substrates with varying affinities for the enzyme. In view of these arguments, a number of workers have developed techniques of measuring the intact glucuronide conjugates. Examples of these are reviewed later in this chapter.

This chapter will be devoted to selected examples of the measurement of metabolites. Many of these methods have been chosen because they also allow the unchanged drug to be measured.

I. Tolmetin and Metabolites

Tolmetin is a nonsteroidal anti-inflammatory drug, which is metabolized by the following pathway:

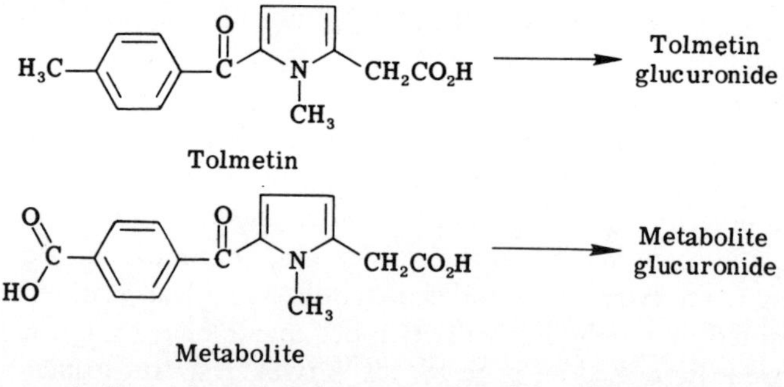

H_3C—C₆H₄—C(=O)—(N-CH_3 pyrrole)—CH_2CO_2H ⟶ Tolmetin glucuronide

Tolmetin

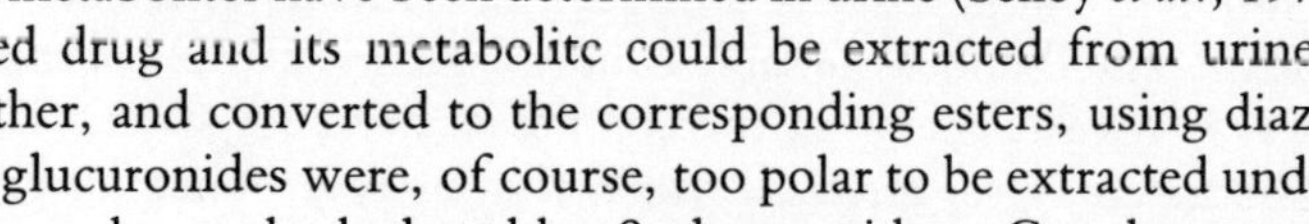

HO(O=)C—C₆H₄—C(=O)—(N-CH_3 pyrrole)—CH_2CO_2H ⟶ Metabolite glucuronide

Metabolite

The unchanged drug alone has been measured in plasma by GLC, while drug and metabolites have been determined in urine (Selley *et al.*, 1974). The unchanged drug and its metabolite could be extracted from urine, using diethyl ether, and converted to the corresponding esters, using diazomethane. The glucuronides were, of course, too polar to be extracted under these conditions and were hydrolyzed by β-glucuronidase. Gas chromatography was carried out on a column of 3% OV-17 on Gas Chrom Q (100–120 mesh) at 230°C. This method illustrates clearly the measurement of a drug and its metabolites. The chloro analog was used as the internal standard and a reasonable separation was obtained.

The method was applied to urine specimens from a number of subjects.

Table 7.1

Percentage of a 300-mg Oral Dose of Tolmetin Excreted in the Urine[a]

Subject	Tolmetin	Tolmetin glucuronide	Metabolite	Metabolite glucuronide
A	17	5	67	5
B	11	58	8	8
C	9	53	4	3

[a] From Selley *et al.* (1974).

Large differences in the amounts excreted were apparent and are shown in Table 7.1.

II. Diethylpropion

The excretion pattern of the appetite suppressant diethylpropion has been studied by Testa and Beckett (1973), and five different metabolites have been identified:

Metabolism takes place by two pathways: dealkylation to yield **II, III, V,** and **VI,** and reduction of the keto group to give **IV, V,** and **VI.** Only about 2% of the orally administered dose appears in the urine unchanged and 85% as metabolites. To measure these, an aliquot of urine is made alkaline and extracted with diethyl ether to remove the amino alcohols **IV, V,** and **VI,** while a further aliquot is treated with phosphate buffer and sodium borohydride to reduce the keto group in metabolites **I, II,** and **III.**

$$C_6H_5-C(=O)-CH(CH_3)-N(R^1)(R^2) \xrightarrow{NaBH_4} C_6H_5-CH(OH)-CH(CH_3)-N(R^1)(R^2)$$

Testa and Beckett went on to separate and measure the diastereoisomers and their enantiomers. However, for a simple separation of the metabolites without resolution of the different isomers, a mixed stationary phase was used: 10% KOH plus 2% Carbowax 20M and 10% Apiezon L on Chromosorb G (AW-DMCS, 100–120 mesh) operated at 200°C. Detection was by FID, and no derivatization was used. A good separation of the unchanged drug and most of its metabolites was obtained with furfurylamphetamine acetate as the internal standard. The keto metabolites **II** and **III** were not completely resolved but could, of course, be converted to the corresponding alcohols and determined by difference. This paper illustrates well the simultaneous determination of a number of very closely related metabolites.

III. Alclofenac

The nonsteroidal, anti-inflammatory alclofenac is eliminated in humans principally as the free drug (A) its glucuronide (AG), dihydroxyalclofenac (DHA), and 3-chloro-4-hydroxyphenylacetic acid (4-HCPA). Roncucci *et al.* (1971) have described a simple method of extracting the metabolites from acidified urine with methyl isobutyl ketone (the glucuronide has to be hydrolyzed first) and derivatization of the evaporated residue with a mixture of HMDS–TMCS in dioxane (2 : 1 : 2). After standing overnight at room temperature, a portion of the mixture is chromatographed on a column of 3% XE-60 on Supelcoport (80–100 mesh) with flame ionization detection. The silylated metabolites can be chromatographed separately (140°C for 4-HCPA, 160°C for A, and 190°C for DHA) or together, starting at 170°C and programing to 210°C. The dihydroxyalclofenac accounts for about 11% of an orally administered dose in humans, and it was postulated to be formed via an intermediate epoxide:

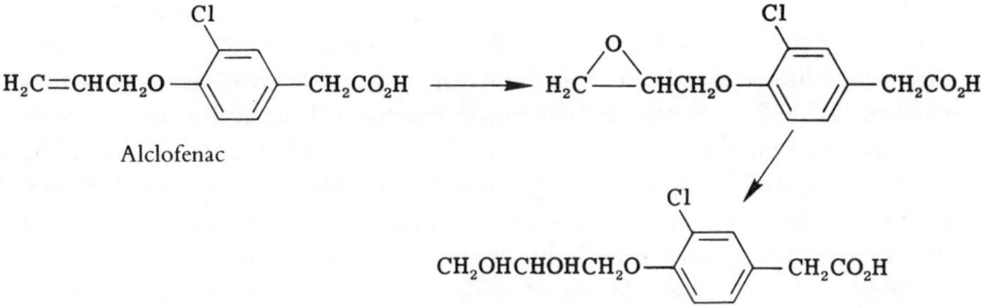

$H_2C{=}CHCH_2O-C_6H_3(Cl)-CH_2CO_2H$ ⟶ $H_2C\overset{O}{—}CHCH_2O-C_6H_3(Cl)-CH_2CO_2H$ ⟶ $CH_2OHCHOHCH_2O-C_6H_3(Cl)-CH_2CO_2H$

Alclofenac

In 1980, Slack and Ford-Hutchinson demonstrated the presence of this epoxide in the urine of healthy volunteers and rheumatoid patients taking 1.5–4 g alclofenac daily by mouth. The authors used GC–MS to identify the metabolite unequivocally, and the study of the five different derivatives used is interesting.

Table 7.2

Derivatization of Alclofenac Epoxide

Reagent	Derivative
1. Diazomethane	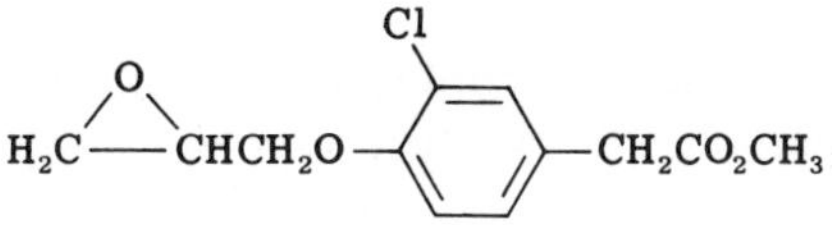 $H_2C\overset{O}{—}CHCH_2O-C_6H_3(Cl)-CH_2CO_2CH_3$
2. BSTFA	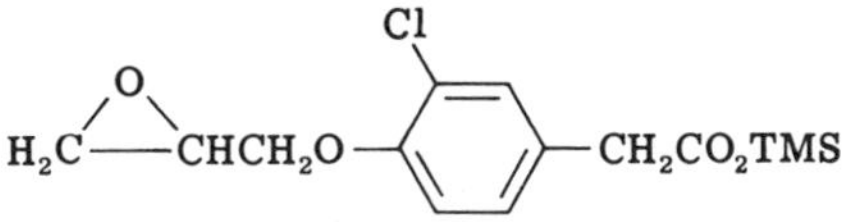 $H_2C\overset{O}{—}CHCH_2O-C_6H_3(Cl)-CH_2CO_2TMS$
3. Diazomethane TMCS BSTFA	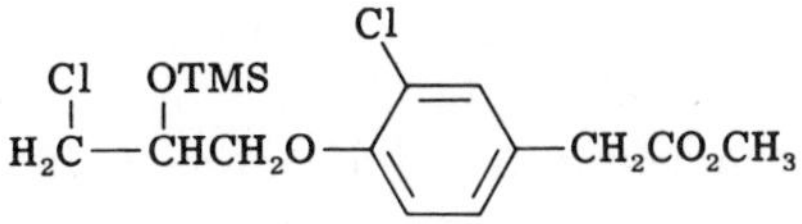 $H_2C(Cl)—CH(OTMS)CH_2O-C_6H_3(Cl)-CH_2CO_2CH_3$
4. Diazomethane *t*-BDMS–imidazole	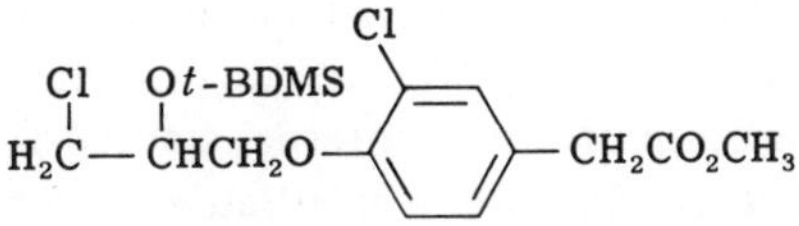 $H_2C(Cl)—CH(O\textit{t}\text{-BDMS})CH_2O-C_6H_3(Cl)-CH_2CO_2CH_3$
5. *t*-BDMS-imidazole	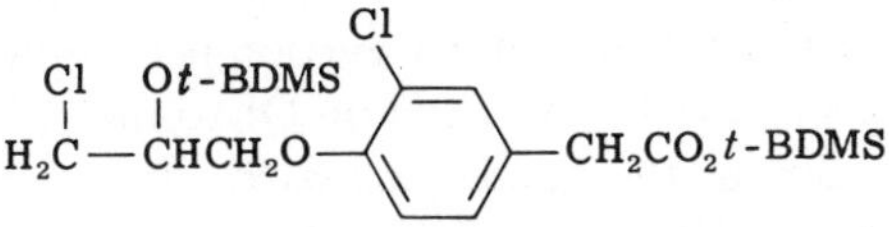 $H_2C(Cl)—CH(O\textit{t}\text{-BDMS})CH_2O-C_6H_3(Cl)-CH_2CO_2\textit{t}\text{-BDMS}$

Urine was acidified to pH 4 using molar HCl, and the epoxide was extracted into chloroform, which was then evaporated to dryness. Derivatization was then carried out, and the nature of the reactions used is summarized in Table 7.2. In the derivatization reactions 3, 4, and 5, the epoxide was first converted to the silyl ether after chlorohydrin formation using the method of Harvey *et al.* (1972). The *t*-BDMS derivatives were found to give a more sensitive and specific response to multiple-ion monitoring, and the structure of the epoxide metabolite was conclusively identified. In order to monitor routinely urine levels of alclofenac epoxide, four of the above derivatives were evaluated, and the diazomethane reaction (1) was found to be most satisfactory. Using this method, approximately 0.005% of the daily dose of alclofenac was found to be excreted in the urine of volunteers and rheumatoid patients. A number of compounds containing the allyl group had previously been shown to be metabolized to the corresponding epoxide, and some of these had been demonstrated to be mutagenic or carcinogenic. Alclofenac has now been withdrawn from the market.

IV. Carbamazepine and Its Epoxide

The epoxide metabolite of the anticonvulsant drug carbamazepine has been shown to be the 10,11-epoxide:

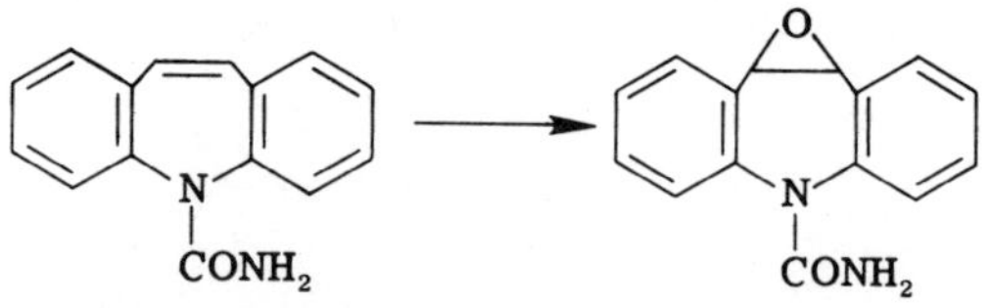

N
$CONH_2$
Carbamazepine

O
N
$CONH_2$

It has been demonstrated to be present in rat and human urine and has been shown to possess pharmacological activity. The suggestion has been made that it should be measured in an attempt to provide a better correlation between blood level and therapeutic effect (Chambers, 1978).

Using imipramine as internal standard, plasma is extracted directly with diethyl ether. Ammonium sulfate is added to the separated ether phase, and after shaking, the ether is decanted and evaporated to dryness. The residue is dissolved in acetone, and a portion is chromatographed on a column of 5% Apiezon L with 0.5% KOH on Diatomite CLQ at 250°C. No mesh size was given. Under the conditions used, the retention times of carbamazepine, epoxide, and internal standard were 4.6, 6.0, and 7.6 min, respectively.

Recoveries were relatively high: 80% for carbamazepine and imipramine and 70% for the epoxide.

Different loadings of KOH were tried, and 0.5% was found to be the most suitable. Increasing the load to 1% resulted in the decomposition of the epoxide. This work illustrates the chromatography of an epoxide metabolite without derivatization.

V. Benzodiazepines and Related Drugs

Many reports have been published on the gas chromatographic separation of benzodiazepines, and one of the earlier papers reports the retention times of metabolites, real and hypothetical, on a 3% OV-17 column without derivatization (Lafargue *et al.*, 1970) (Table 7.3). In the following year Zingales (1971) described the measurement of the desmethyl and lactam metabolites of chlordiazepoxide.

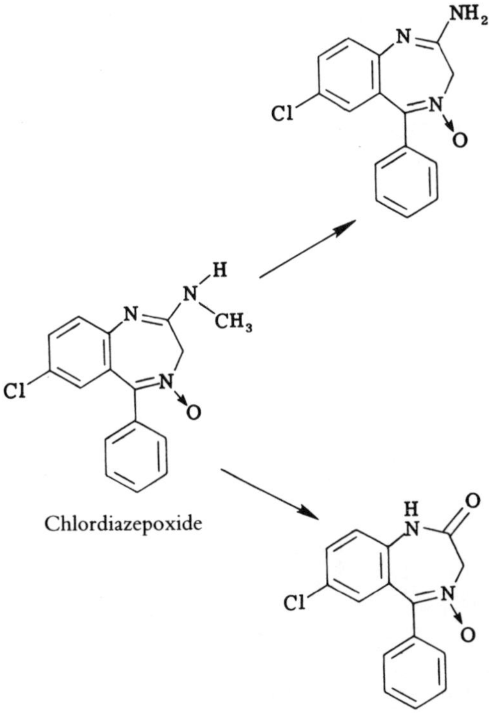

Chlordiazepoxide

Trebbi *et al.* (1975) describe the measurement of pinazepam and its three metabolites in serum, urine, and brain using ECD. The metabolic pathway

Table 7.3

Retention Times of Benzodiazepine Metabolites on 3% OV-17 at 250°C[a]

Compound	Retention time (min)
Desmethyltetrazepam	13.5
Desmethyldiazepam	15.4
Desmethylchlordiazepoxide	15.6
7-Aminonitrazepam	41.0
7-Acetamidonitrazepam	123.8

[a] From Lafargue *et al.* (1970).

for pinazepam is shown in Fig. 7.1. Serum is extracted at alkaline pH with diethyl ether and the drug and its metabolites are back-extracted into 2 *M* HCl. After washing the acid phase with diethyl ether, it is made alkaline (pH 9), and the drug and its metabolites are extracted into fresh diethyl ether. The solvent is evaporated, the residue redissolved in hexane–acetone (4 : 1), and a portion is injected onto a column of 3% GE XE-60 on Gas Chrom Q (60–80 mesh) at 250°C. The sensitivity of the method for pinazepam in serum is about 5–10 ng/ml and 15–20 ng/ml for the metabolites.

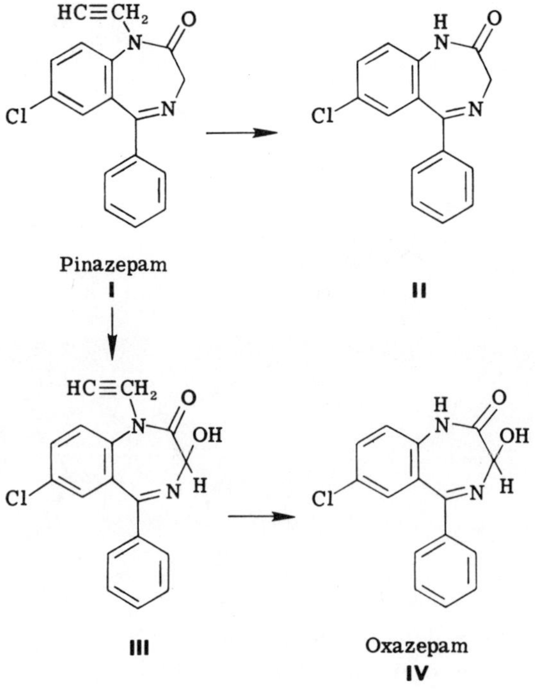

FIG. 7.1. Metabolic pathway for pinazepam.

Recoveries are relatively high, ranging from 76% for metabolite **III** to 84% for the unchanged drug. Using this method, pinazepam and metabolite **II** were detected in the serum and brain of rat, dog, and man, while all four compounds were detected in rat urine.

Puglisi *et al.* (1976) were able to measure the drug 7-iodo-1,3-dihydro-1-methyl-5-(2′-fluorophenyl)-2*H*-1,4-benzodiazepin-2-one in blood and urine using ECD. The drug could be measured at a concentration of 2 ng/ml, while the desmethyl metabolite could be measured down to 4 ng/ml.

Loxapine is an antipsychotic drug belonging to the dibenzoxazepine class, and its metabolism, together with that of the closely related antidepressant amoxapine is illustrated in Fig. 7.2. The gas chromatographic analysis of these drugs and their metabolites has been described by Cooper and Kelly (1979). Secondary amines are derivatized, using TFAA, while phenolic groups are converted to the corresponding TMS ethers. In this way both the unchanged drugs and their metabolites can be measured. The 8-methoxy-loxapine analog is used as an internal standard, and serum samples are made alkaline (pH 9.7) before extraction with ethyl acetate. Drugs and metabolites are back-extracted into 0.1 *M* HCl, which is washed with fresh ethyl acetate before being made basic. Drugs and metabolites are extracted into ethyl acetate, and after evaporation, the relevant derivatization is carried out.

To determine the drugs and their metabolites in urine, a preliminary enzymatic hydrolysis was carried out to split any conjugates. After being made alkaline, the hydrolysate mixture was extracted with ethyl acetate. Back-extraction into 0.1 *M* HCl was carried out, and the aqueous phase was lyophilized before derivatization. Chromatography was carried out on a column of 3% SP-2100 on Supelcoport (100–120 mesh) at 255°C. Urine levels of drugs and metabolites were sufficiently high to be determined by FID, while ECD was used for serum concentrations. The good separation achieved under these conditions is illustrated in Fig. 7.3. Recoveries ranged from 70% for loxapine to 90% for 8-hydroxyloxapine. The method was sensitive enough to allow the measurement of concentrations of 5 ng/ml of 8-hydroxyloxapine and 8-hydroxyamoxapine.

The antianxiety agent clobazam is metabolized mainly by N-demethylation to desmethylclobazam:

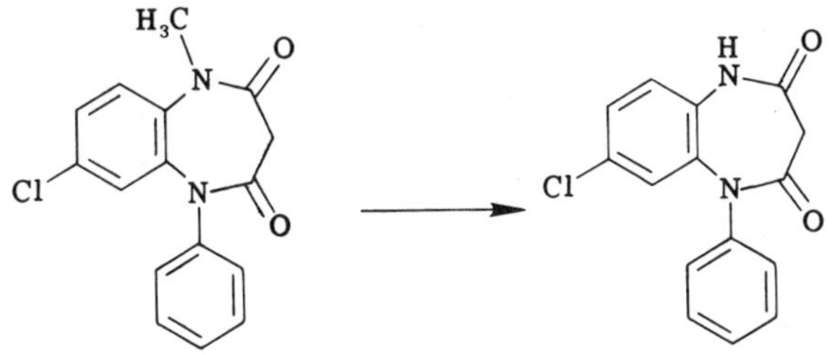

H_3C N O Cl N O → H N O Cl N O

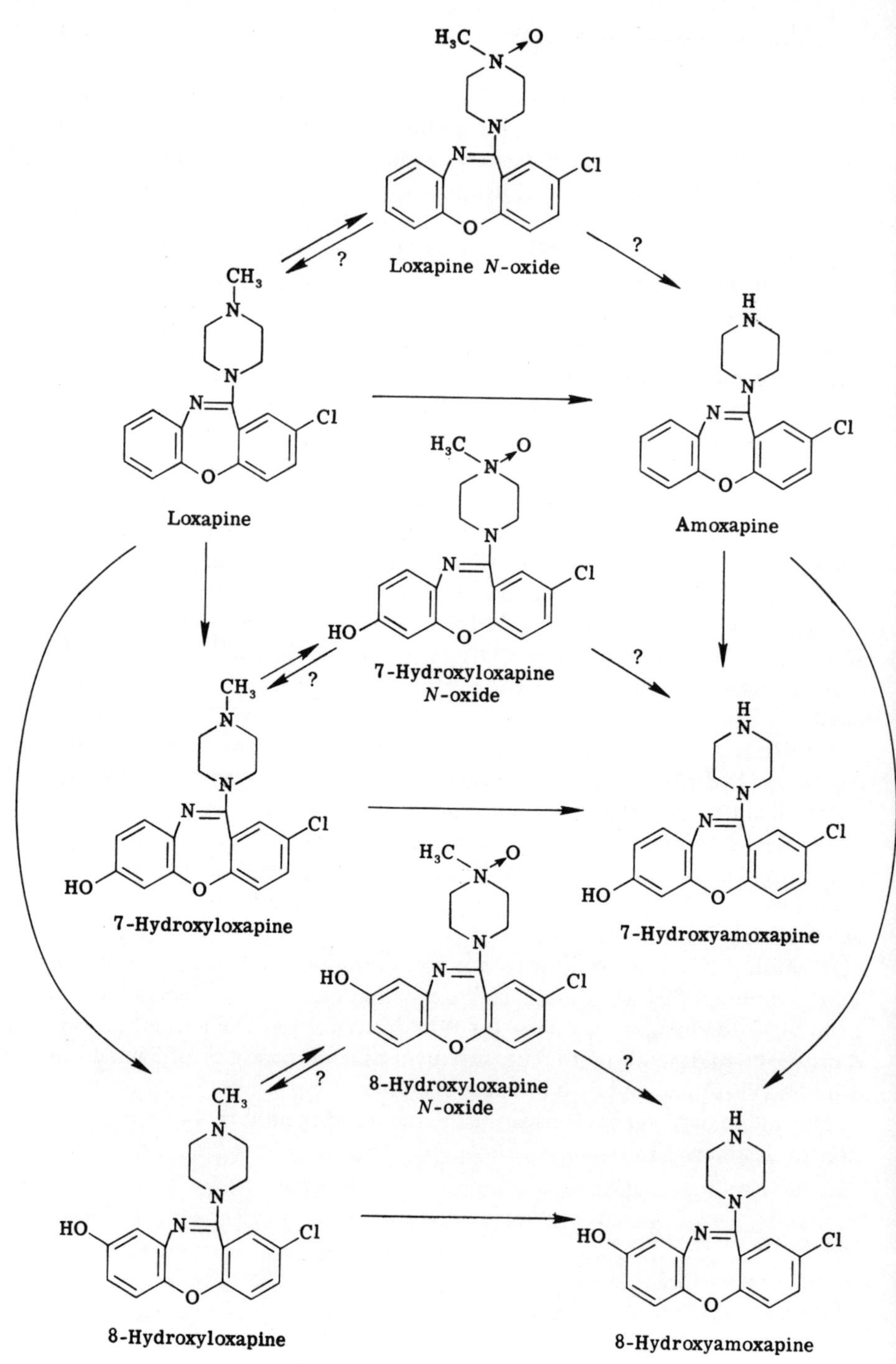

FIG. 7.2 Metabolism of loxapine.

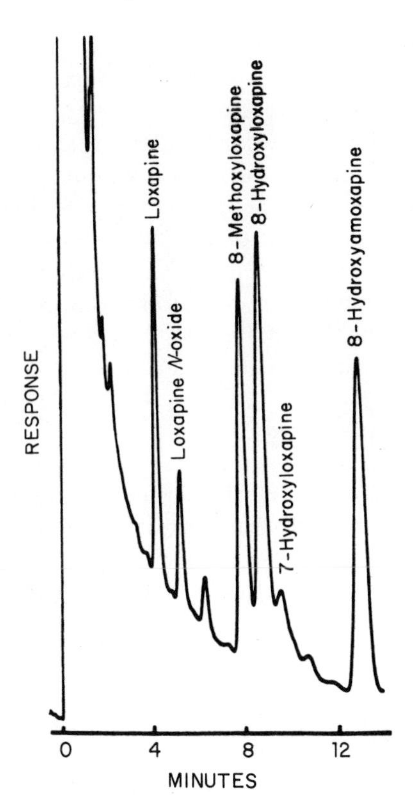

FIG. 7.3 Chromatograph showing separation of loxapine and its metabolites. From Cooper and Kelly (1979). Reproduced with permission of the copyright owner.

This metabolite also possesses pharmacological activity, and the simultaneous determination of both has been reported by Caccia *et al.* (1979). In this simple method, diazepam is added to plasma as internal standard, and then the plasma is buffered at pH 9.5 and extracted with benzene. The organic phase is separated, evaporated to dryness, and redissolved in acetone for gas chromatography. A column of 5% OV-25 on Chromosorb G AW-DMCS (80–100 mesh) was used at the relatively high temperature of 290°C, just below the maximum operating temperature for this phase. Under the conditions used by the authors, the retention times of diazepam, clobazam, and its metabolite were 2, 3, and 5 min, respectively. Electron-capture detection was used, and recoveries were high: 94–98% for clobazam and 90–94% for the desmethyl metabolite. The authors measured plasma levels of clobazam and its metabolite in guinea pigs after intraperitoneal (ip) injection of the drug. In later papers Caccia *et al.* (1980a,b) studied tissue levels of drug and metabolite in the mouse and rat.

Greenblatt (1980) measured plasma levels in man, using a method as simple as that of Caccia *et al.* but with a different stationary phase. Plasma was extracted, without pH adjustment, using benzene containing 1.5% isopentanol to minimize adsorption to glassware. Diazepam was added as internal standard. The organic phase was evaporated to dryness, and the residue was dissolved in toluene containing 15% isopentanol. Injections were made onto a column packed with 10% OV-101 on Chromosorb W HP (80–100 mesh) and operated at 265°C. Electron-capture detection was used, and on each working day the column was primed with 2–3 μl azolectin (1 mg/ml) in benzene. Under the above conditions the retention times of diazepam, clobazam, and desmethylclobazam were 5.5, 7.4, and 8.6 min, respectively.

A more polar stationary phase, OV-17, was tried, but this was found to produce an asymmetrical peak with desmethylclobazam, and the nonpolar OV-101 was chosen. Recoveries of all three compounds were greater than 95%, and precision was good. The limit of detection was about 3–5 ng/ml for clobazam and 5–10 ng/ml for the metabolite. Plasma levels of both were measured in a volunteer after a single oral dose of 20 mg of clobazam.

VI. Probenecid

The metabolism of probenecid has been studied using gas chromatography (Conway and Melethil, 1975). It is metabolized in humans by dealkylation and oxidation:

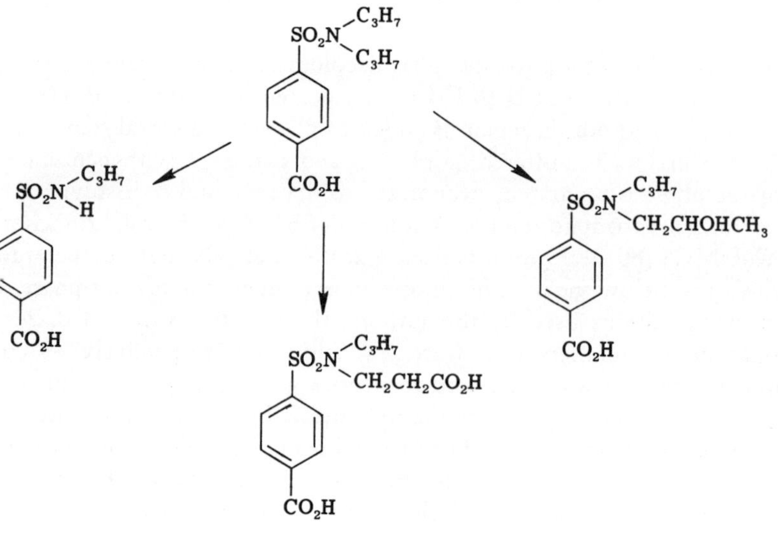

$SO_2N(C_3H_7)_2$ / CO_2H

$SO_2N(C_3H_7)(H)$ / CO_2H

$SO_2N(C_3H_7)(CH_2CH_2CO_2H)$ / CO_2H

$SO_2N(C_3H_7)(CH_2CHOHCH_3)$ / CO_2H

Conjugates were hydrolyzed by heating urine specimens in 5 *M* HCl for 4 h at 100°C, and the free acids were extracted into methylene chloride. Esterification was carried out using diazopropane, and then the internal standard *N,N*-dibenzyl-2,5-dimethylbenzenesulfonamide was added. This compound does not contain a carboxylic acid group. Chromatography was then carried out on a 6-ft stainless steel column packed with 10% OV-1 on Chromosorb W-HP (80–100 mesh) at 250°C, using flame ionization detection. An example of the separation obtained is shown in Fig. 7.4.

The retention times are given in Table 7.4 along with the mean recoveries from urine, which were relatively high. Although the limit of detection was not given, the authors claimed that their method was capable of measuring unchanged drug and metabolites in urine for at least 3 days following a single 500-mg oral dose. Interestingly, the primary alcohol, presumably the precursor of the carboxylic acid metabolite, was not detected. This method is a good example of the simple measurement of a number of metabolites of different polarities under a single set of GC conditions.

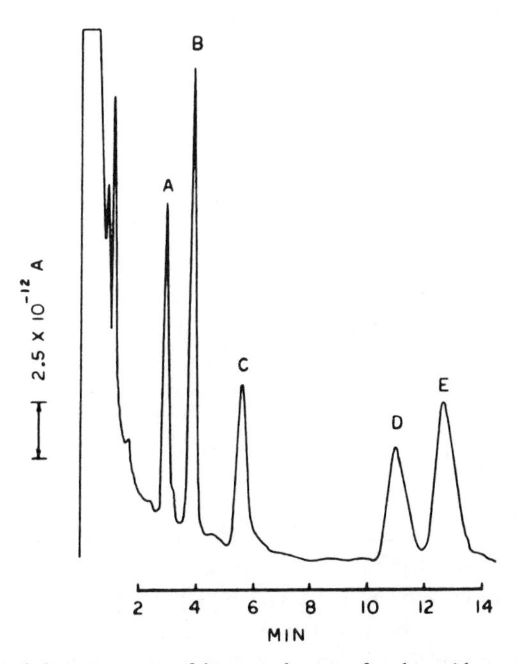

FIG. 7.4. A typical chromatogram of the propyl esters of probenecid metabolites excreted in human urine: A, mono-*n*-propyl metabolite; B, probenecid; C, secondary alcohol metabolite; D, carboxylic acid metabolite; E, internal standard *N,N*-dibenzyl-2,5-dimethylbenzenesulfonamide. From Conway and Melethil (1975).

Table 7.4

Retention Times of Probenecid and Its Metabolites[a]

Compound	t_R (min)	Recovery from urine (%)
Probenecid	3.9	98
Mono-*n*-propyl metabolite	3.1	83
sec-Alcohol metabolite	5.7	94
Carboxylic acid metabolite	11.1	85
Internal standard	12.8	—

[a] From Conway and Melethil (1975).

VII. Chloral Hydrate and Its Metabolites

The metabolism of chloral hydrate has been studied by Breimer *et al.* (1974) who identified trichloroethanol, its glucuronide, and trichloroacetic acid as metabolites in blood:

$$CCl_3CH(OH)_2 \longrightarrow CCl_3CH_2OH \longrightarrow CCl_3CO_2H$$

$$CCl_3CH_2OH \downarrow \text{Glucuronide}$$

These compounds are relatively volatile and were measured in blood and urine by head space analysis. Samples were split into two portions, one for the determination of unconjugated trichloroethanol and the other for chloral hydrate, total trichloroethanol, and trichloroacetic acid.

A. Free Trichloroethanol

Whole blood in a vial was treated with lead acetate to prevent the *in vitro* conversion of chloral hydrate to trichloroethanol. The contents of the vial were then homogenized, and the vial was placed in a water bath at 60°C. After equilibrium had been reached (3 h), a measured amount of vapor phase was injected onto a 1.8-m × 3-mm glass column packed with 10% OV-17 on Gas Chrom Q (80–100 mesh) at 125°C. Electron-capture detection was used.

B. Chloral Hydrate, Total Trichloroethanol, and Trichloroacetic Acid

A 2-ml sample of whole blood is added to a vial containing 1 ml concentrated sulfuric acid. The strong acid not only hydrolyzes the glucuronide conjugate to give free trichloroethanol but also, like lead acetate, prevents the *in vitro* conversion of chloral hydrate to trichloroethanol by erythrocytes. The vial is then treated in the same way as in the measurement of free trichloroethanol. A portion of the vapor phase is injected onto the same OV-17 column, and chloral hydrate and total trichloroethanol are determined. Dimethyl sulfate (0.1 ml) is then added to the vial to convert the trichloroacetic acid to its methyl ester. A period of 4 h is allowed to elapse to reestablish equilibrium, and then a portion of the vapor phase is injected onto the GC column. The chromatographic separation achieved is shown in Fig. 7.5. Head-space analysis is extremely useful in the analysis of chloral hydrate and its metabolites, particularly since the compounds are relatively polar. The sample handling time is short, and (if one discounts the 3–7-h period necessary for equilibration) the method is rapid.

The authors investigated a number of stationary phases, including some containing orthophosphoric and terephthalic acids, and found a 10% loading of OV-17 to be the most satisfactory. The detection limits reported were not particularly low, considering that electron-capture detection was used: chloral hydrate and trichloroethanol, 500 ng/ml, and trichloroacetic acid, 100 ng/ml, from blood and urine. Blood levels were measured after the iv administration of chloral hydrate (60 mg/kg) to a dog. The authors conclude their paper by suggesting that their method might be suitable for the study of the metabolism of trichloroethylene.

VIII. Metoprolol Metabolites

The β-adrenoceptor blocking drug metoprolol is metabolized by oxidative pathways affecting both side chains (Fig. 7.6). About 5% of an oral dose is excreted unchanged and 65 and 10% as the pharmacologically inactive metabolites H 117/04 and H 104/83, respectively. The pharmacologically active H 119/66 and H 105/22 account for about 10 and 0.1%, respectively. Neither the unchanged drug nor its metabolites are excreted as conjugates. No method has been published for the measurement of the carboxylic acid metabolite H 104/83, presumably because it is inactive and accounts for only a small percentage of the dose. The others are all extractable from alkaline urine, using a mixture of diethyl ether–dichloromethane

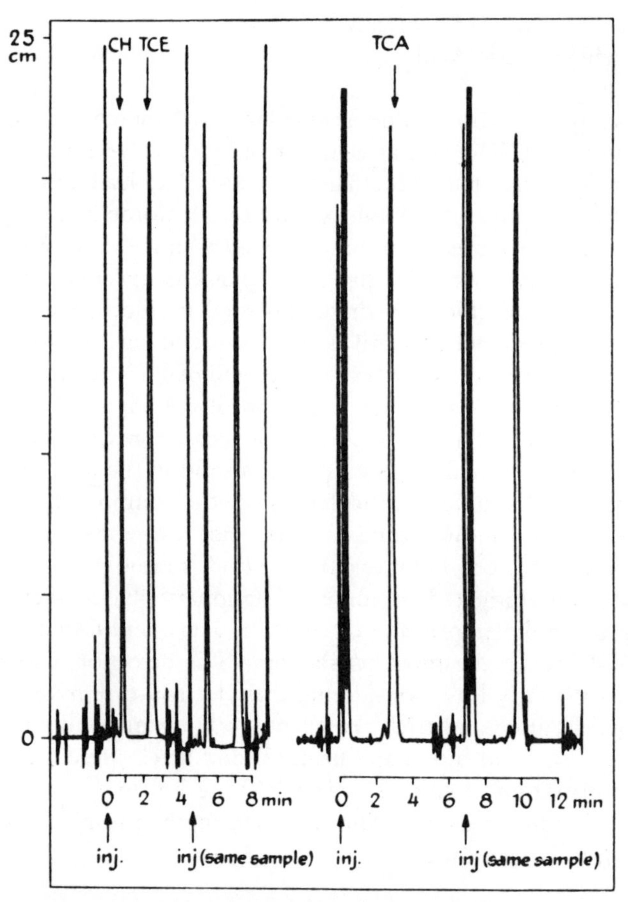

FIG. 7.5. Gas chromatograms of chloral hydrate (CH), trichloroethanol (TCE), and trichloroacetic acid methyl ester (TCA) after direct injection of head-space gas, in equilibrium with a 2.0-ml blood sample + 1.0 ml of concentrated sulfuric acid. Blood concentrations of the compounds were: CH, 10.2 μg/ml; TCE, 12.6 μg/ml (injection volume 1000 μl); and TCA, 10.5 μg/ml (injection/μl). From Breimer *et al.* (1974).

(3:2). The α-hydroxy metabolite H 119/66 was derivatized, using heptafluorobutyrylimidazole in benzene, (1:3) at room temperature for 1 h.

The sample was then evaporated, and the residue was dissolved in toluene and washed with *M* NaOH before injection of a portion of the toluene phase onto a 6-ft × 2-mm ID glass column packed with 3% OV-1 on Gas Chrom Q (100–120 mesh) and operated at 165–180°C (Quarterman *et al.*, 1980). The β-hydroxy metabolite H 105/22 was derivatized with a mixture of trifluoroacetic anhydride in toluene (2:5) for 1.5 h at room temperature.

$$\text{OCH}_2\text{CH(OH)CH}_2\text{NHCH(CH}_3)_2 \;-\; \text{C}_6\text{H}_4 \;-\; \text{CH(OH)CH}_2\text{OCH}_3 \xrightarrow{\text{HFBIm-Bz}} \text{OCH}_2\text{CH(OCOC}_3\text{F}_7)\text{CH}_2\text{N(COC}_3\text{F}_7)\text{CH(CH}_3)_2 \;-\; \text{C}_6\text{H}_4 \;-\; \text{CH(OCOC}_3\text{F}_7)\text{CH}_2\text{OCH}_3$$

The reaction mixture was then evaporated to dryness, and the residue was dissolved in toluene ready for injection onto a similar size column packed with 3% OV-17 on Gas Chrom Q (80–100 mesh) operated at 180–195°C.

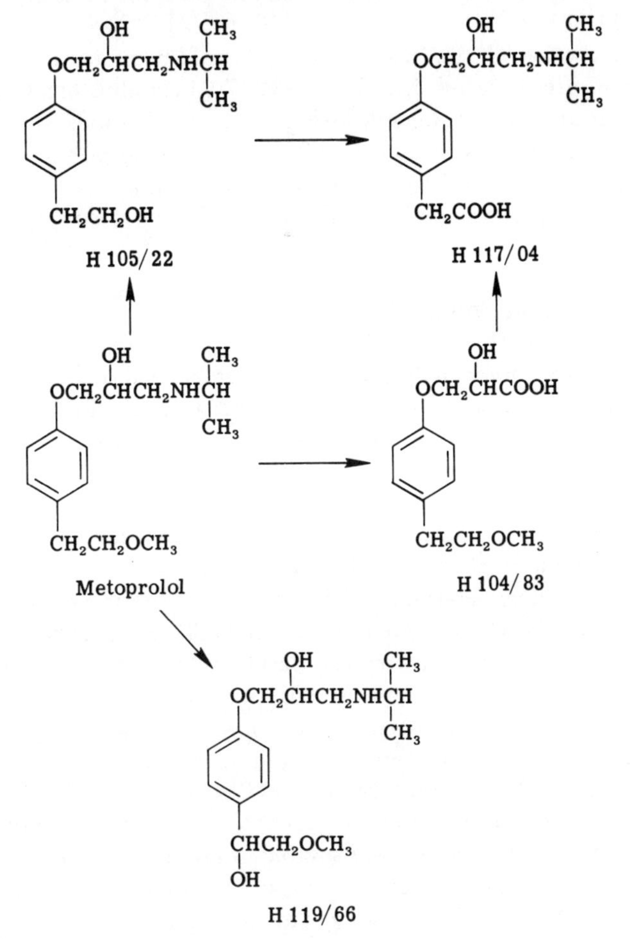

FIG. 7.6. Metabolism of β-adrenoceptor blocking drug metoprolol.

The major, but pharmacologically inactive, metabolite H 117/04 in urine was measured after the following derivatization:

$$\text{HOOCCH}_2\text{-C}_6\text{H}_4\text{-OCH}_2\text{CH(OH)CH}_2\text{NHCH(CH}_3)_2 \xrightarrow[\text{2. CF}_3\text{CH}_2\text{OH}]{\text{1. HFBIm}} \text{CF}_3\text{CH}_2\text{O}_2\text{CCH}_2\text{-C}_6\text{H}_4\text{-OCH}_2\text{CH(OCOC}_3\text{F}_7)\text{CH}_2\text{N(COC}_3\text{F}_7)\text{CH(CH}_3)_2$$

Gas chromatography was carried out at 180°C on the same column as that used for H 119/66.

The metabolites H 119/66 and H 105/22 could be measured at concentrations of 6 ng free base/ml. Metabolite H 117/04 could be measured in urine at a concentration of 500 ng/ml. Using the above methods, the recovery of all the metabolites was greater than 90%. Depending on the metabolite to be measured and the biological fluid used, different internal standards were used: alprenolol, oxprenolol, or the ethyl analog of metoprolol H 93/47.

IX. Phenothiazine Metabolites

Phenothiazines are usually extensively metabolized, often to a variety of different metabolites including sulfoxides and sulfones. A typical example is thioridazine (Fig. 7.7). Fluorimetry is sensitive but measures total drug, i.e., thioridazine and metabolites. Dinovo *et al.* (1976) have developed a simple extraction procedure that allows the quantitative estimation of the unchanged drug and its metabolites without derivatization, using FID.

Plasma is made alkaline with NaOH, and the drug and its metabolites are extracted into a heptane–toluene mixture, 4 : 1. The compounds to be measured are then back-extracted into 0.1 *M* HCl, and part of this aqueous phase is removed for fluorimetric estimation, while the remainder is basified and shaken with a small volume (100 μl) of heptane–toluene containing the internal standard chlorpromazine. A portion of this organic phase is chromatographed on a 6-ft × 2-mm ID column packed with 3% OV-17 on Chromosorb Q (100–120 mesh) at 275°C. The disulfone and disulfoxide metabolites of thioridazine were too polar to be extracted into the heptane–toluene solution but were readily soluble in methylene chloride.

The retention times under the chromatographic conditions described above are given in Table 7.5 along with the mean recovery from plasma. Using FID, a minimum detectability of 300 ng/ml from 4 ml of plasma was

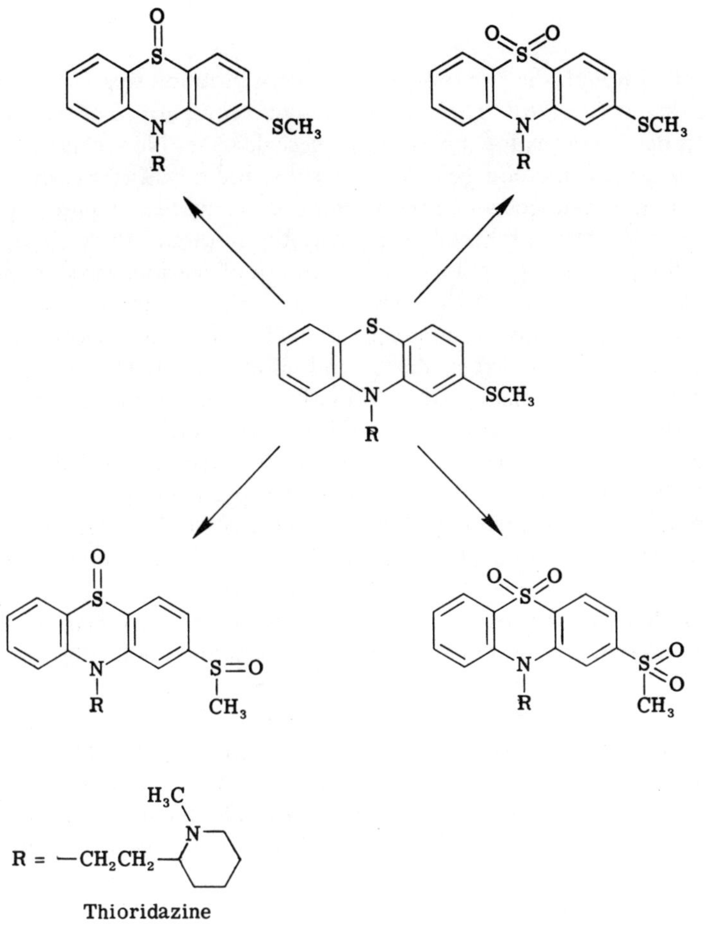

R = $—CH_2CH_2—$ (N-methylpiperidin-2-yl)

Thioridazine

FIG. 7.7 Metabolism of thioridazine.

Table 7.5

Retention Time and Recovery from Plasma of Thioridazine and Metabolites[a]

	t_R (min)	Recovery (%)
Thioridazine	3.6	77
Thioridazine (*S*)-sulfoxide (mesoridazine)	8.3	56
Thioridazine (*R*)-sulfoxide	12.8	38
Northioridazine	3.8	—
Northioridazine (*S*)-sulfoxide	8.7	—
Thioridazine disulfone	22.8	—
Thioridazine disulfoxide	20.9	—
Unknown impurity	9.1	—

[a] From Dinovo *et al.* (1976).

reported, although the incorporation of a concentration step was claimed to reduce this to 50 ng/ml. The authors present a comparison of the GLC and fluorescence methods for the assay of mesoridazine. Not surprisingly, the chromatographic method gives lower results, and the difference increases in those patient specimens when the amount of metabolites would be greater. The separation of a number of other phenothiazines, including chlorpromazine, trifluoperazine, prochlorperazine, and fluphenazine, are shown.

Vanderheeren *et al.* (1976) describe the chromatographic analysis of thioridazine metabolites in human plasma. This paper also reports the determination of unchanged thioridazine and another phenothiazine, perazine. The authors confine themselves to only the metabolites of thioridazine that are pharmacologically active: mesoridazine and sulforidazine. Plasma is made alkaline with 10 *M* NaOH, and the specimens are heated in a water bath at 100°C for 15 min. After cooling, the specimens are extracted three times with heptane containing 1.5% isoamyl alcohol. The drugs and metabolites are then back-extracted into 0.1 *M* HCl, and the aqueous phase is made alkaline with more 10 *M* NaOH. Three extractions with the heptane–isoamyl alcohol mixture are carried out, and after centrifugation, the organic phase is evaporated to dryness, and the residue is redissolved in a very small volume (10 μl) of heptane containing 1.5% isoamyl alcohol. Chromatography is carried out on a 1.83-m × 4-mm ID glass column packed with 3% XE-60 on Gas Chrom Q (80–100 mesh) operated at 240°C for 15 min and then to 270°C at a rate of 10°/min. Perazine was used as the internal standard for thioridazine and its metabolites. The retention times were as follows: perazine 0.4 min, thioridazine 2.5 min, mesoridazine 5.8 min, and sulforidazine 6.4 min. A chromatogram obtained from an extract of human plasma is shown in Fig. 7.8.

Recoveries for the unchanged drug and its metabolites were all greater than 97%, and the limit of detection from 5 ml plasma was 100 ng/ml for thioridazine and 150 ng/ml for the two metabolites. This paper is a good example of the use of temperature programing to allow metabolites to be determined along with unchanged drug.

The measurement of loxapine, amoxapine, and metabolites has already been discussed on page 165. Perphenazine and its sulfoxide have been measured following derivatization (Larson and Naestoft, 1975). A single extraction into toluene followed by derivatization with BSA was sufficient to measure unchanged perphenazine. A second extraction was necessary for the sulfoxide metabolite and the compound used as the internal standard, 8-chloroperphenazine (Fig. 7.9). Gas chromatography was carried out on a 1.5-m × 4-mm ID glass column packed with 1% OV-17 on Celite JJCQ (100–120 mesh) operated at 305°C. Both unchanged drug and metabolite could be measured down to 0.2 ng/ml.

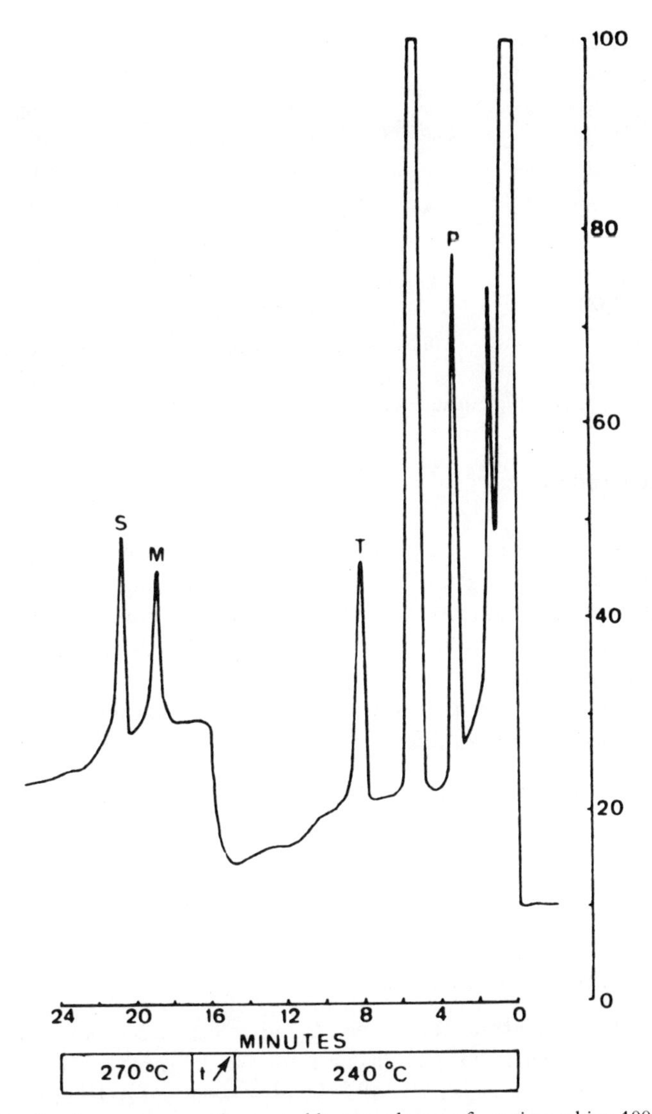

FIG. 7.8. Gas chromatogram of extracted human plasma of a patient taking 100 mg thioridazine four times daily. Plasma taken 12 h after the last dose. P, Perazine (internal standard, 1 μg/ml); T, thioridazine (0.94 μg/ml); M, mesoridazine (1.36 μg/ml); S, sulforidazine (0.64 μg/ml). From Vanderheeren *et al.* (1976).

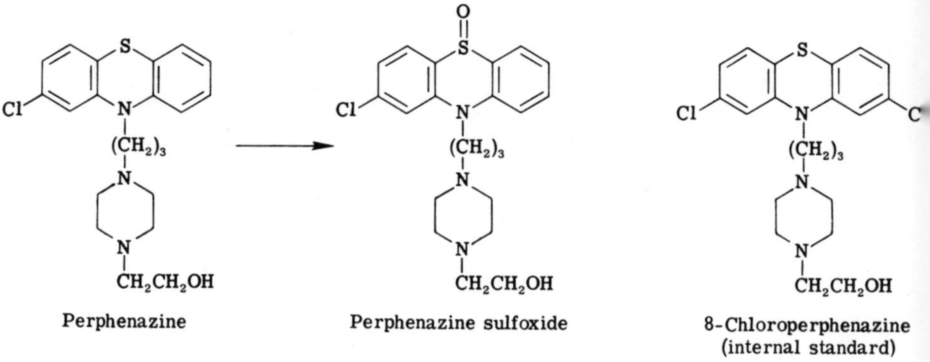

Perphenazine → Perphenazine sulfoxide

8-Chloroperphenazine (internal standard)

FIG. 7.9. Perphenazine and its sulfoxide and the internal standard used in its GC detection.

X. Glucuronides

Many drugs or their metabolites are excreted by the kidney as highly polar glucuronide conjugates. Although these conjugates can be hydrolyzed chemically or enzymatically and the resulting aglycone measured in the normal way, a number of workers have measured the intact glucuronide by gas chromatography. Apart from the intellectual and practical challenges of measuring glucuronides in this way, there are a number of reasons why hydrolysis may be undesirable:

1. if strong acid is used, the resulting aglycone may have its structure modified or may be destroyed;
2. if strong acid is used to hydrolyze the conjugate, there is no proof that the conjugate was in fact a glucuronide since sulfate, glycine, and other conjugates can also be split in this way;
3. the presence of enzyme inhibitors in biological specimens can make quantitative hydrolysis difficult.

The glucuronic acid moiety of the conjugate has two types of group that need to be derivatized: the carboxylic acid and the three hydroxyl groups. The carboxylic-acid group is usually simply converted to the methyl ester, and then the alcoholic groups are acetylated, trimethylsilylated, or methylated. Derivatization of all four groups to give a pertrimethylsilyl derivative has also been reported.

In 1974 Ehrsson *et al.* described the gas chromatography of ether glucuronides as methyltrifluoroacetyl derivatives. As model compounds the authors used the glucuronide conjugates of 1-naphthol and androsterone, which were purchased as pure compounds. Esterification was carried out

Table 7.6

Retention Times for Glucuronide Conjugates[a]

Compound	t_R (min), OV-1	t_R (min), OV-17
1-Naphthol	0.3, at 170°C	0.4 at 200°C
Me-TFA derivative of 1-naphthol glucuronide	6.15 at 170°C	2.4
Me-TFA androsterone glucuronide	6.0 at 250°C	—

[a] From Ehrsson *et al.* (1974).

using diazomethane in diethyl ether, which was allowed to react with the conjugate (100 μg in methanol) for 15 min at 20°C. The solution was then evaporated to dryness under a stream of nitrogen, and the residue was mixed with ethyl acetate (200 μl) and trifluoroacetic anhydride (25 μl) and vortexed for 15 min at 20°C. In this reaction mixture the derivatized conjugate was stable for at least 24 h at 20°C; however, attempts to remove the remaining trifluoroacetic anhydride by evaporation or solvent extraction resulted in decomposition. The derivatized conjugates were chromatographed on a 6-ft × 2-mm ID glass column packed with either 1% OV-1 on Chromosorb W (60–80 mesh) or 2% OV-17 on Chromosorb W HP (80–100 mesh). The OV-1 column was operated at 170 and the OV-17 at 200°C. Flame ionization detection was used, of course, since the use of electron capture was impossible without removal of the unreacted trifluoroacetic anhydride. The retention times for the two derivatized conjugates are given in Table 7.6.

In the following year Marcucci *et al.* (1975) examined the glucuronides of oxazepam and loxazepam in urine. The glucuronides were removed from urine by passage through a column packed with Amberlite XAD-2 and then washing the column with 0.01 *M* formic acid. This washing was reported to improve the recovery of the glucuronides. Elution of the glucuronides from the column was achieved by using 80% aqueous acetone, and the eluate was dried under a vacuum. The residue was then redissolved in a little methanol ready for derivatization.

The derivatization was carried out in two steps: (1) reaction with ethereal diazomethane to esterify the carboxylic acid function after removal of solvent, and (2) silylation with HMDS and TFAA in pyridine (5 : 1 : 10). The diazomethane also methylates the N-1 nitrogen (Fig. 7.10). The reaction was complete after 15 min, and the derivatives were chromatographed on a column of 3% OV-17 on Gas Chrom Q (100–120 mesh). No column temperature was given.

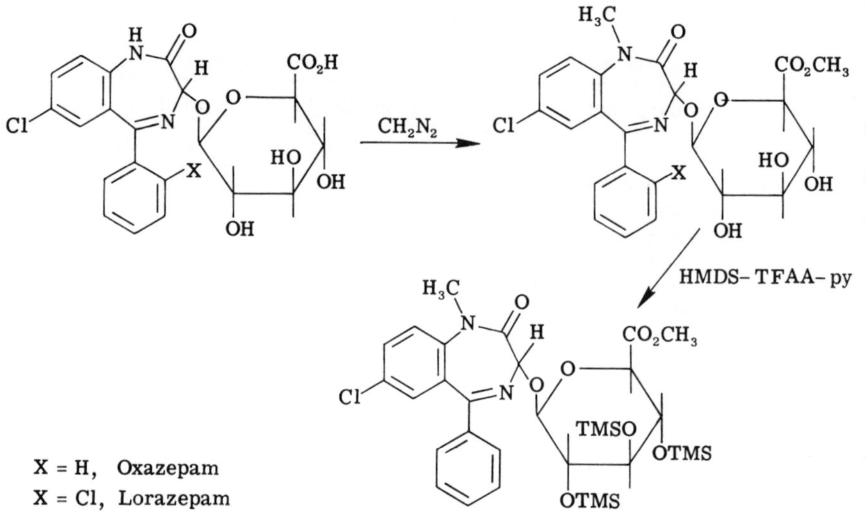

$$\text{Oxazepam/Lorazepam glucuronide} \xrightarrow{CH_2N_2} \text{N-methyl methyl ester} \xrightarrow{\text{HMDS-TFAA-py}} \text{TMS derivative}$$

X = H, Oxazepam
X = Cl, Lorazepam

FIG. 7.10. Derivatization of oxazepam and loxazepam glucuronides.

Using this method, oxazepam glucuronide was measured in rabbit, guinea pig, and human urine but was only found in trace amounts in rat urine. Such a reaction scheme is not applicable when N-1-methyl-C-3-hydroxy metabolites are present in urine along with N-1-desmethyl-C-3-hydroxy metabolites (e.g., *N*-methyloxazepam glucuronide with oxazepam glucuronide) since the same derivative is formed. In such a situation diazopropane should be used.

In 1976, Thompson and Gerber used GC–MS to characterize the glucuronides of 1- and 2-naphthols, 2-, 3- and 4-hydroxybiphenyls, and *m*- and *p*-hydroxyphenylphenylhydantoins. The glucuronides were collected by perfusing an isolated rat liver with the appropriate compound and collecting the bile. Small aliquots of bile were evaporated to dryness and permethylated using methylsulfinylmethide carbanion and methyl iodide as shown in Fig. 7.11. The permethylated glucuronides could be chromatographed on a number of columns: 1 or 3% SE-30 and 1% OV-17, all on Gas Chrom Q (100–120 mesh) and 5% OV-17 on Gas Chrom P (80–100 mesh) at 80°C and then programing to 100°C at 2°C/min.

In addition to these derivatives, a number of aryl glucuronic acid conjugates have been examined, using mass spectrometry of the acylated methyl esters. These derivatives were formed using acetic anhydride and methanesulfonic acid, extracting into chloroform, and evaporating to dryness. The residue was then dissolved in methanol and esterified with diazomethane.

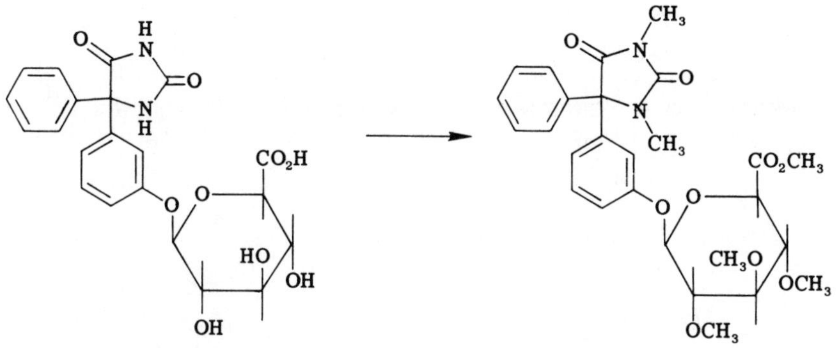

m-Hydroxyphenylphenyl-
hydantoin glucuronide

FIG. 7.11. Permethylation of m-hydroxyphenylphenylhydantion glucuronide.

No gas chromatographic separation of the derivatives was reported (Paulson *et al.*, 1973).

The glucuronide conjugate of the β-blocker befunolol was extracted from biological fluids, using ion-exchange chromatography, and derivatized using BSTFA (Kawahara and Ofuji, 1979). The derivative was chromatographed on a 1.5% OV-17 column at 230°C.

In the methods described here the glucuronides are generally removed from urine by adsorbing them on to XAD-2 resin and then eluted with a suitable solvent. This procedure is not readily adaptable to large numbers of specimens unless disposable columns are used. A much more versatile method could be achieved if it were possible to extract the intact glucuronides from urine or even plasma into an organic solvent. Glucuronides, by their very nature, are too polar for direct extraction but Kammer and Goldzieher (1969) reported the removal of steroid glucuronides from urine as ion pairs with pyridinium sulfate. These authors also summarized previous attempts to remove intact glucuronides from biological fluids using charcoal, a derivative of Sephadex, and even egg albumin.

XI. Narcotics

A. Propoxyphene and Metabolites

The analgesic α-d-propoxyphene is metabolized in humans to norpropoxyphene:

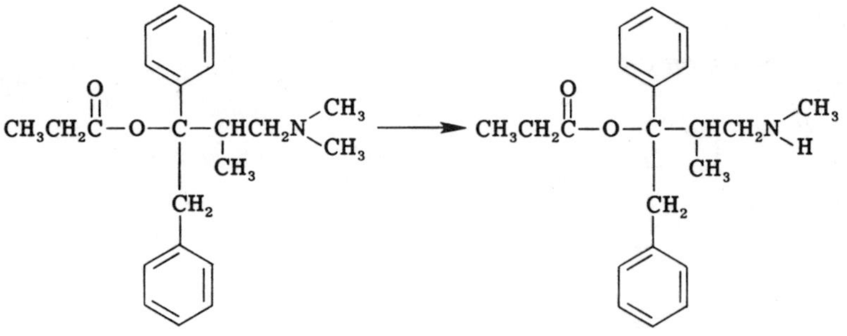

$CH_3CH_2\overset{O}{\overset{\|}{C}}-O-\underset{CH_2C_6H_5}{\overset{C_6H_5}{C}}-\underset{CH_3}{C}HCH_2N(CH_3)_2 \longrightarrow CH_3CH_2\overset{O}{\overset{\|}{C}}-O-\underset{CH_2C_6H_5}{\overset{C_6H_5}{C}}-\underset{CH_3}{C}HCH_2N(CH_3)H$

Propoxyphene

The analysis of the unchanged drug was, in the beginning, difficult because of decomposition during chromatography. Sparacino *et al.* (1973) carried out a detailed investigation and demonstrated that the decomposition was not thermal but was due to the acidic surface activity of the solid support. High performance supports such as Chromosorb W HP and Gas Chrom Q were found to be best.

Verebely and Inturrisi (1973) were able to measure both unchanged drug and metabolite in plasma and urine. Plasma was made alkaline with a carbonate–bicarbonate buffer (pH 9.8), and after adding a single drop of octyl alcohol, extraction with butyl chloride was carried out. A back-extraction into 0.2 *M* HCl was carried out, and the acidic aqueous phase was washed with *n*-hexane, which was discarded. After making the aqueous phase alkaline with 60% NaOH, the drug and its metabolite were extracted into chloroform, which was evaporated, and the residue was redissolved in carbon tetrachloride for gas chromatography using a column of 3% SE-30 on Gas Chrom Q (80–100 mesh) at 216°C. The drug and its metabolite were extracted directly from alkaline urine into chloroform. An internal standard (SKF 525-A) was added before extraction of plasma and urine.

The estimation of norpropoxyphene is interesting because the metabolite is converted via a cyclic intermediate to the corresponding amide at a pH $>$ 11. This occurs on addition of the 60% NaOH.

This transformation was first described and studied by McMahon *et al.* (1971). Interestingly, if norpropoxyphene itself is chromatographed under the conditions described, three peaks are produced. However, conversion to the amide results in a chromatographically stable compound. Using this method, 10 ng of propoxyphene and 50 ng of norpropoxyphene could be detected per ml plasma, starting with an initial volume of 4 ml. The authors were able to use their method to demonstrate that on chronic dosing metabolite levels were three times that of the drug itself.

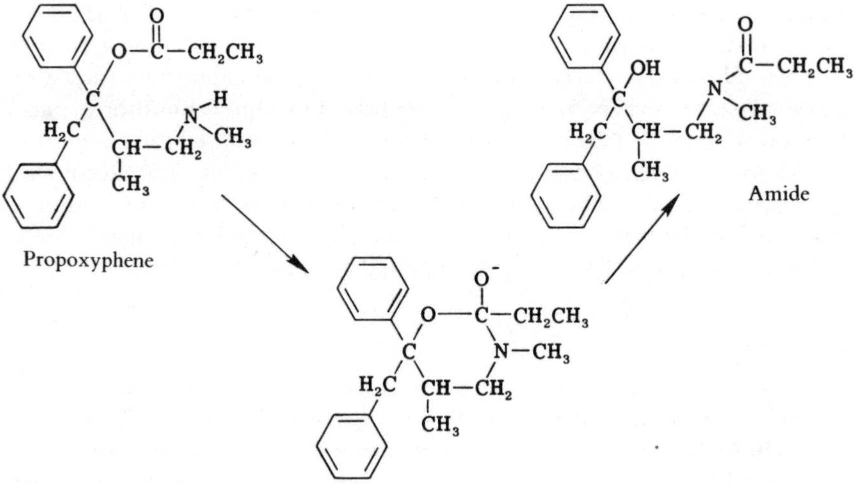

Propoxyphene

Amide

B. Methadone

The narcotic analgesic methadone undergoes ring closure followed by demethylation during its metabolism. Both metabolites and the unchanged drug have been separated on a 6-ft × 2-mm ID column of 3% SE-30 (Inturrisi & Verebely, 1972). An adaptation of the Wolen and Gruber (1968) propoxyphene method is used to extract unchanged drug, metabolites, and the internal standard SKF 525A. Plasma pH is adjusted to 9.8, and extrac-

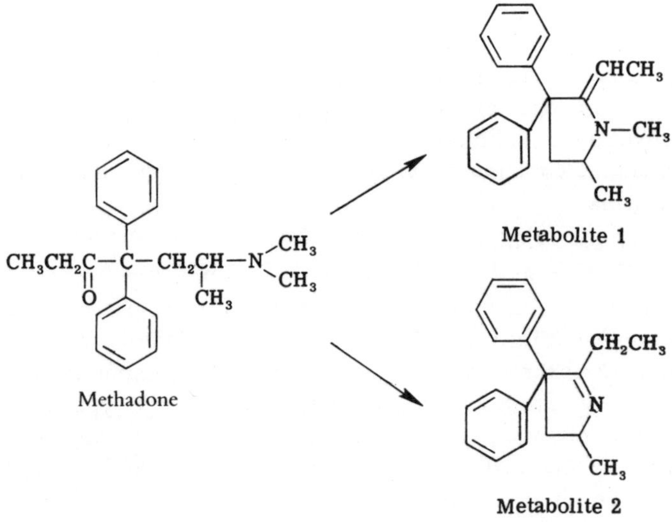

Methadone

Metabolite 1

Metabolite 2

tion is carried out with *n*-butyl chloride. Back-extraction into 0.2 *M* HCl is performed, and the aqueous phase is separated and made alkaline with 60% NaOH. The final extraction step is carried out using chloroform, and after evaporation to dryness, the residue is dissolved in a little chloroform, and a portion is injected. A column temperature of 200°C is used for plasma and 180°C for urine extracts. The detection system used is FID. Recoveries greater than 95% are obtained, and the retention times under the conditions described by the authors are as follows: metabolite 2, 2.3 min; metabolite 1, 3.2 min; methadone, 4.7 min; and internal standard, 8.8 min.

C. Acetylmethadol

Studies of opiate dependence showed that acetylmethadol given orally three times a week was as effective in suppressing narcotic withdrawal as daily methadone. This was in accord with the finding that its activity lasted three times longer than methadone. Acetylmethadol is metabolized according to the scheme shown in Fig. 7.12.

Billings *et al.* (1973) were able to measure the metabolite noracetylmethadol in the plasma and tissues of the rat and identified dinoracetylmethadol as a second metabolite in that species. Plasma or tissue homogenate was made alkaline (pH 9.5) with *M* NaOH and extracted with butyl chloride. The organic phase was separated and evaporated to dryness, and the residue was dissolved in toluene; then 50 μl of 1% trichloroacetyl chloride in toluene was added, and derivatization was allowed to take place at 70–80°C for 15 min. The reaction mixture was evaporated to dryness, and the residue was dissolved in hexane for injection. Chromatography was carried out on a 2-ft × 2.5-mm ID column packed with 3% OV-1 on Gas Chrom Q (100–120 mesh) operated at 205°C. Because of the nature of the derivative formed, electron-capture detection was employed; no internal standard appeared to be used. Although no limit of detection was given, concentrations of noracetylmethadol were reported down to 100 ng/g in brain, liver, and lung, as well as plasma.

Kaiko *et al.* (1975) measured noracetylmethadol and dinoracetylmethadol after converting them to the corresponding amides:

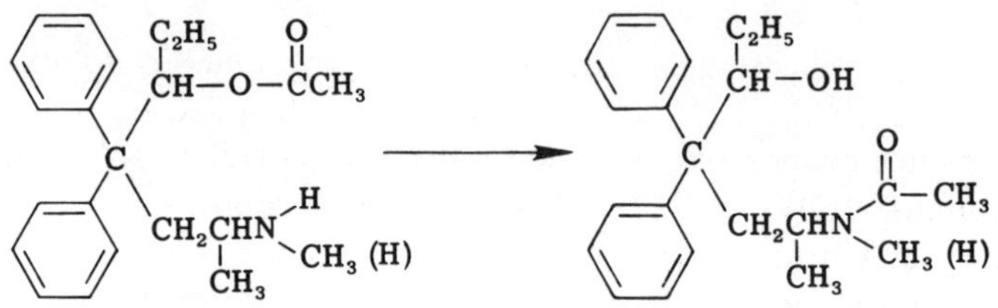

C_2H_5 — $CH{-}O{-}\overset{O}{\overset{\|}{C}}CH_3$; C ; $CH_2CHN(H)CH_3$ (H) ; CH_3

→

C_2H_5 — $CH{-}OH$; C ; $CH_2CHN(\overset{O}{\overset{\|}{C}}{-}CH_3)CH_3$ (H) ; CH_3

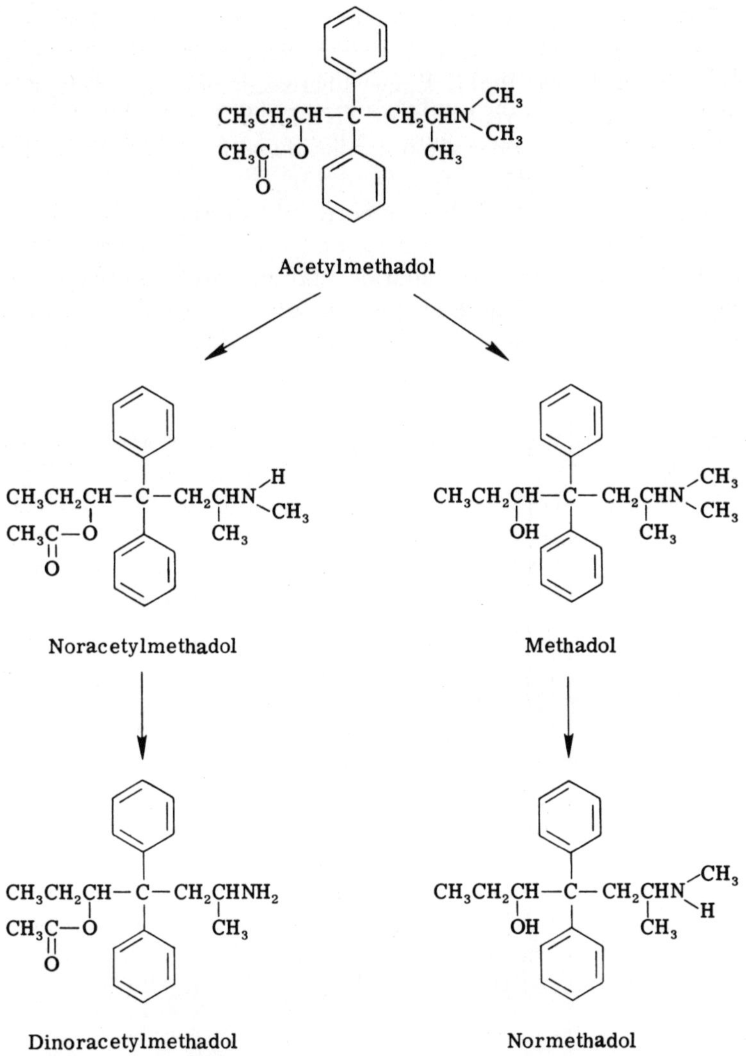

Acetylmethadol

Noracetylmethadol

Methadol

Dinoracetylmethadol

Normethadol

FIG. 7.12. Metabolism of acetylmethadol.

This is done by heating an alkaline (pH > 13) aqueous solution of the metabolites at 70°C for 30 min. The amides are then extracted into chloroform, which is evaporated, and the residue is dissolved in fresh chloroform for chromatography. A 6-ft × 2-mm ID glass column packed with 3% SE-30 on Gas Chrom Q (80–100 mesh) is used and is operated at 235°C. Flame ionization detection is used, and the authors claim that the method is capable of measuring acetylmethadol and dinoracetylmethadol down to 10 ng/ml and noracetylmethadol to 25 ng/ml. Under these chromatographic conditions a good separation is achieved between unchanged drug, the two metabolites, and the internal standard SKF 525A (Fig. 7.13.)

This rearrangement under alkaline conditions is very similar to that described for norpropoxyphene (pp. 182–183).

Lau and Henderson (1976) have reported the measurement of acetylmethadol and its metabolites in patients on methadone maintenance. Alkaline plasma is extracted with butyl chloride, and the separated organic phase is evaporated to dryness. The residue is redissolved in acetone, and a portion of this solution is injected onto a 4-ft × 1/8-in ID glass column packed with

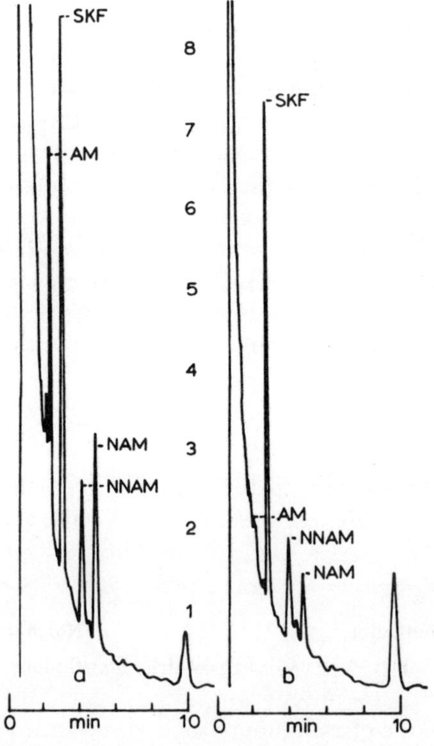

FIG. 7.13. Chromatograms of human plasma extracts from a patient who received an oral dose of 50 mg of acetylmethadol (AM). The internal standard SKF 525-A was added directly to the plasma and the extract prepared. Noracetylmethadol (NAM) and dinoracetylmethadol (NNAM) present in the extract were quantitatively converted to their corresponding amides just prior to the final step of the extraction procedure. a, 4-h Post-drug plasma; b, 48-h post-drug plasma. Retention times are AM, 1.8 min; SKF, 2.5 min; NNAM, 3.6 min; NAM, 4.2 min. From Lau and Henderson (1976).

3% OV-25 on Gas Chrom Q (100–120 mesh) operated at 170°C with flame ionization detection. This procedure is carried out to allow the measurement of unchanged acetylmethadol and methadone. After this has been done, the acetone is evaporated, and the residue is treated with 0.25 m*M* trichloroacetyl chloride in toluene. After 2 h at 70°C, the mixture is evaporated to dryness, and the residue is dissolved in hexane for gas chromatography using a 3-ft × 1/8-in ID glass column packed with 3% OV-17 on Gas Chrom Q (100–120 mesh). The column is operated at 235°C, and electron-capture detection is used.

This paper by Lau and Henderson is particularly interesting because it contains a detailed comparison of a number of derivatizing agents: trichloroacetyl chloride, pentafluorobenzoyl chloride, heptafluorobutyryl chloride, heptafluorobutyric anhydride, and trifluoroacetic anhydride. Irrespective of the reagent chosen, the optimum reaction time was found to be 2 h and the optimum concentration 0.25 m*M*. Quantitative conversion was claimed for the three metabolites noracetylmethadol, dinoracetylmethadol, and normethadol, but the derivatization of methadol was unsuccessful with the conditions and reagents discussed. Trichloroacetyl chloride and heptafluorobutyryl chloride produced fewer side products, and trichloroacetyl chloride was proposed as the reagent of choice for the measurement of the relevant metabolites in plasma and urine extracts. Interestingly, pyridine and triethylamine were described as being "inadequate" catalysts: pyridine because it was very difficult to remove by evaporation and triethylamine because it did not affect the yield of any of the reactions studied.

D. Diacetylmorphine

Not surprisingly, diacetylmorphine and its metabolite have been the subject of a number of studies, and derivatization is necessary prior to chromatography. In most species diacetylmorphine is rapidly deacetylated to 6-acetylmorphine:

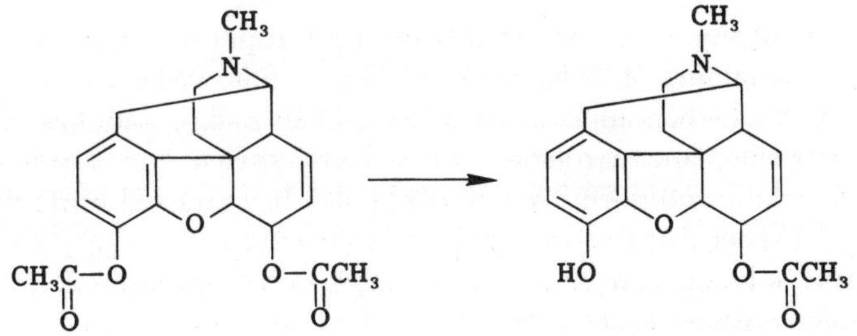

CH_3 N $CH_3C(=O)-O$ … O … $O-C(=O)CH_3$ → CH_3 N HO … O … $O-C(=O)CH_3$

The first method we shall consider is that of Smith and Cole (1975), which uses N-FID. Ethylmorphine acetate was added to blood as the internal standard, and the pH was adjusted to 9.0 using a glycine–NaOH buffer. Extraction was carried out using benzene, and the separated organic phase was evaporated to dryness. Derivatization was carried out using a 1 : 5 mixture of trifluoroacetic anhydride and benzene and letting the reaction take place at 50°C for 1 h. Portions of the reaction mixture or ethyl acetate dilutions were injected onto a 213-cm × 0.4-cm ID glass column packed with 2% OV-17 on Diatomite C-AW (100–120 mesh) operated at a temperature of 250°C. The hydrogen flow rate to the detector was ~24 ml/min and was optimized daily.

Using the above conditions, the retention times were as follows: monoacetylmorphine TFA, 2.1 min; ethylmorphine acetate, 4.1 min; and diacetylmorphine, 6.7 min. The extraction procedure was deliberately made as short as possible to minimize the possible decomposition of diacetylmorphine. The authors reported that this decomposition was appreciable *in vitro* and estimated it to be 8–15 nmol/ml/min in blood. Quantitation was reported down to 100 ng/ml, with detection as low as 20 ng/ml. The authors administered a 16-mg dose of diacetylmorphine intravenously to a 25-kg Irish greyhound and were able to monitor the rapid disappearance of the parent drug and metabolite. The derivatization was judged necessary in order to reduce adsorption of the metabolite during chromatography.

In humans, diacetylmorphine is excreted in the urine mainly as conjugated morphine with a little free morphine. Small amounts of diacetylmorphine and 6-acetyl morphine have been detected in the urine after large intravenous doses. In a detailed study Yeh *et al.* (1977) devised a method of detecting these metabolites and separating those compounds that could possibly be metabolites: morphine ethereal sulfate, free and conjugated normorphine, morphine 6-glucuronide, and dihydromorphinone. As volunteers, the study employed healthy post-addict federal prisoners. Conjugates were split either with 20% HCl in an autoclave for 30 min or by using bacterial β-glucuronidase at pH 6.8 and 37°C. A 4-h incubation was used for morphine 6-glucuronide and 18 h for morphine and normorphine conjugates.

6-Acetylmorphine was extracted from urine at pH 8–8.5, using ethylene dichloride containing 30% isopropanol. There followed a back-extraction into 0.1 *M* HCl, then adjustment of the pH to 8–8.5, addition of 1 g of sodium chloride, and extraction with fresh ethylene dichloride–isopropanol. Morphine and normorphine could also be extracted using the same procedure, except that the urine pH was adjusted to 10. On evaporation of the organic solvent, conversion to the silyl, acetyl, propionyl, or trifluoroacetyl derivatives could be carried out. The authors reported the retention

Table 7.7

Retention Times of Derivatives of Diacetylmorphine Metabolites[a]

	TFAA (210°C) (min)	TMS (220°C) (min)	TMS (230°C) (min)
6-Acetylmorphine	6.9	—	7.1
Codeine	5.8	—	—
Morphine	3.8	7.4	4.3
Norcodeine	9.1	—	—
Normorphine	6.2	8.8	—

[a] From Yeh *et al.* (1977).

times of the TFAA and TMS derivatives (Table 7.7). The chromatography was carried out using a 0.9-m × 2-mm glass column packed with 3% OV-17 on Gas Chrom Q (60–80 mesh).

XII. Local Anesthetics and Their Metabolites

Cocaine is extensively metabolized to benzoylecgonine and ecgonine according to the following scheme:

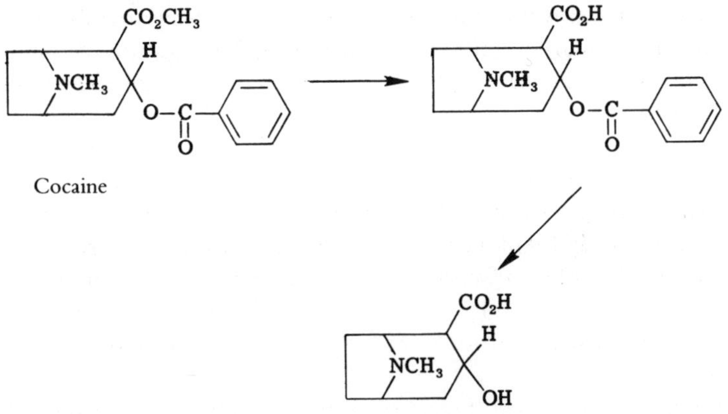

Cocaine

The unchanged drug can be determined in plasma and urine following extraction into cyclohexane (Blake *et al.*, 1974). In order to measure low levels of cocaine, reduction with lithium aluminum hydride is first carried out, followed by conversion to the diheptafluorobutyrate (page 190).

Javaid and colleagues (1975) applied the above method to measure the metabolites too. However, a careful reading of their paper reveals that no extraction of the metabolites from plasma or urine was described: the authors simply state that the cyclohexane extraction for cocaine did not remove ecgonine and benzoylecgonine. They do, however, state that both metabolites can be derivatized using a mixture of hexafluoroisopropanol and heptafluorobutyric anhydride (1 : 2) at 75°C for 30 min. The mixture was then evaporated to dryness under a stream of nitrogen, and the residue was dissolved in cyclohexane.

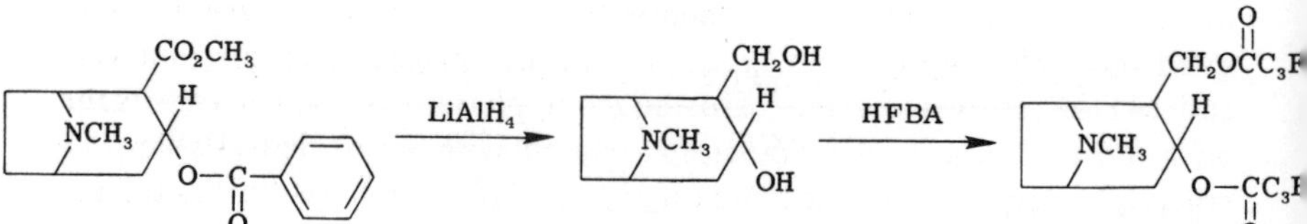

Using a 6-ft × 2-mm ID column packed with 5% OV-1 on GH P (80–100 mesh), the ecgonine derivative had a retention time of 12.5 min at a column temperature of 130°C and a nitrogen flow rate of 30 ml/min. The benzoylecgonine derivative had a retention time of 15.6 min at 190°C with a flow rate of 50 ml/min. The reduced and derivatized cocaine was chromatographed on the same column but at a much lower temperature: it had a retention time of 4.2 min at 110°C and a flow rate of 20 ml/min.

In the following year Nation *et al.* (1976) described the extraction from plasma and quantitation of lidocaine and its dealkylated metabolite, monoethylglycinexylidide (MEGX), without derivatization (Fig. 7.14). Plasma was made alkaline with 10 *M* NaOH, and the unchanged drug and

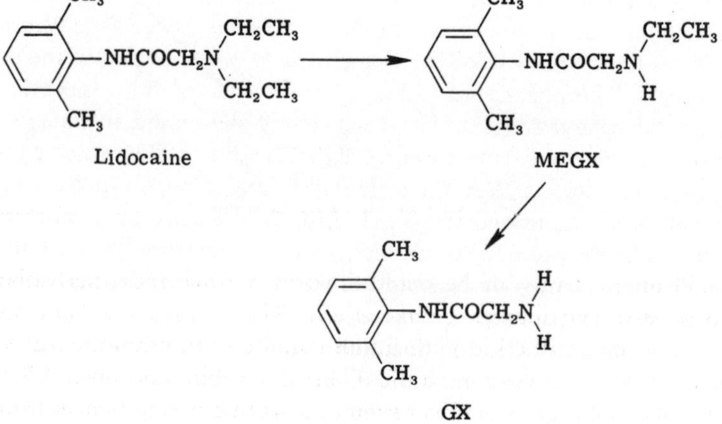

FIG. 7.14. Dealkylation of lidocaine.

metabolite were extracted into diethyl ether. The organic phase was separated and shaken with *M* HCl, and the aqueous phase, containing lidocaine and MEGX, was separated. This aqueous phase was then made alkaline with more sodium hydroxide and extracted with fresh diethyl ether. The ether was evaporated in a water bath at 40°C, and then the tube was plunged into ice to condense the remaining ether vapor. This procedure was repeated until approximately 10 μl remained, and this was injected onto a 2.5-ft × $\frac{1}{4}$-in OD glass column packed with 2% Ucon 75-H-90,000 + 2% KOH on 80–100-mesh Gas Chrom Q at 175°C. Flame ionization was used as the means of detection, and benzhexol hydrochloride was used as the internal standard. Under the conditions described, the retention times of lidocaine, MEGX, and internal standard were 4.8, 7.6, and 10 min, respectively.

The authors did not claim a limit of detection, but the lowest point on their published calibration curve was 0.05 μg. The use of 2% KOH minimized drug adsorption and increased peak symmetry. The authors compared their results with those of Strong and Atkinson (1972) who used a mixed phase of 3% SE-30 and 3% OV-17 (6 : 1). The latter obtained asymmetric peaks for the metabolites, and the MEGX calibration curve was nonlinear below 0.5 μg/ml.

In an earlier study, Adjepon-Yamoah and Prescott (1974) were able to measure lidocaine, MEGX, and the didesethyl metabolite glycylxylidide (GX) following acetylation of the metabolites with an acetic anhydride–pyridine mixture. Using a nitrogen-selective flame ionization detector, 10–30 ng/ml could be measured. A 4-ft × $\frac{1}{4}$-in OD glass column packed with 3% cyclohexane dimethanol succinate on 100–120-mesh Gas Chrom Q was used with temperature programing from 200 to 245°C.

Finally, Rosseel and Bogaert (1978) reported the separation of lidocaine and metabolites by the use of a capillary column. Trimecaine was added to plasma as the internal standard along with 40 μl of triethylamine to reduce adsorption. The plasma was acidified and extracted with methylene chloride, which was discarded. The plasma was then made alkaline using 5 *M* NaOH and extracted with fresh methylene chloride. The organic phase was evaporated to dryness, and the residue was treated with a mixture of hexane and trifluoroacetic anhydride (5 : 1). After 5 min at room temperature, the mixture was evaporated to dryness, and the tubes were stored at −18°C. Gas chromatography was carried out within 6 h when the derivatized compounds were redissolved in ethyl acetate. The column was 20 m × 0.5 mm ID, coated with OV-17, and operated at 190°C. The retention times were as follows: GX TFA, 2.8 min; MEGX TFA, 3.3 min; lidocaine, 3.8 min; and trimecaine, 5.6 min. The lower limit of detection was quoted as being about 20 ng/ml.

References

Adjepon-Yamoah, K. K., and Prescott, L. F. (1974). *J. Pharm. Pharmacol.* **26,** 889.
Billings, R. E., Booher, R., Smits, S., Pohland, A., and McMahon, R. E. (1973). *J. Med. Chem.* **16,** 305.
Blake, J. W., Ray, R. S., Noonan, J. S., and Murdick, P. W. (1974). *Anal. Chem.* **46,** 288.
Breimer, D. D., Ketelaars, H. C. J., and Van Rossum, J. M. (1974). *J. Chromatogr.* **88,** 55.
Caccia, S., Dallabio, M., Guiso, G., and Zanini, M. G. (1979). *J. Chromatogr.* **164,** 100.
Caccia, S., Guiro, G., Samanin, R., and Garattini, S. (1980a). *J. Pharm. Pharmacol.* **32,** 101.
Caccià, S., Guiro, G., and Garattini, S. (1980b). *J. Pharm. Pharmacol.* **32,** 295.
Chambers, R. E. (1978). *J. Chromatogr.* **154,** 272.
Conway, W. D., and Melethil, S. (1975). *J. Chromatogr.* **115,** 222.
Cooper, T. B., and Kelly, R. G. (1979). *J. Pharm. Sci.* **68,** 216.
Curry, S. H. (1980). "Drug Disposition and Pharmacokinetics," 3rd Ed. Blackwell, Oxford.
Dinovo, E. C., Gottschalk, L. A., Nandi, B. R., and Geddes, P. G. (1976). *J. Pharm. Sci.* **65,** 667.
Ehrsson, H., Walle, T., and Winkström, S. (1974). *J. Chromatogr.* **101,** 206.
Greenblatt, D. J. (1980). *J. Pharm. Sci.* **69,** 1351.
Harvey, D. J., Glazener, L., Stratton, C., Johnson, D. B., Hill, R. M., Horning, E. C., and Horning, M. G. (1972). *Res. Commun. Chem. Pathol. Pharmacol.* **4,** 247.
Inturrisi, C. E., and Verebely, K. (1972). *J. Chromatogr.* **65,** 361.
Javaid, J. I., Dekirmenjian, H., Brunngraber, E. G., and Davis, J. M. (1975). *J. Chromatogr.* **110,** 141.
Kaiko, R. F., Chatterjie, N., and Inturrisi, C. E. (1975). *J. Chromatogr.* **109,** 247.
Kammer, C. S., and Goldzieher, J. W. (1969). *Anal. Biochem.* **28,** 492.
Kawahara, K., and Ofuji, T. (1979). *J. Chromatogr.* **168,** 266.
Lafargue, P., Pont, P., and Meunier, J. (1970). *Ann. Pharm. Fr.* **28,** 477.
Larson, N.-E., and Naestoft, J. (1975). *J. Chromatogr.* **109,** 259.
Lau, D. H., and Henderson, G. L. (1976). *J. Chromatogr.* **129,** 329.
Marcucci, F., Bianchi, R., Airoldi, L., Solmona, M., Fanelli, R., Chiabrando, C., Frigerio, A., Mussini, E., and Garattini, S. (1975). *J. Chromatogr.* **107,** 285.
McMahon, R. E., Ridolfo, A. S., Culp, H. W., Wolen, R. L., and Marshall, F. J. (1971). *Toxicol. Appl. Pharmacol.* **19,** 427.
Nation, R. L., Triggs, E. J., and Selig, M. (1976). *J. Chromatogr.* **116,** 188.
Parke, D. V. (1968). "The Biochemistry of Foreign Compounds." Pergamon Press, Oxford.
Paulson, G. D., Zaylskie, R. G., and Dockter, M. M. (1973). *Anal. Chem.* **45,** 21.
Puglisi, C. V., de Silva, J. A. F., and Leon, A. S. (1976). *J. Chromatogr.* **118,** 371.
Quarterman, C. P., Kendall, M. J., and Jack, D. B. (1980). *J. Chromatogr.* **183,** 92.
Roncucci, R., Simon, M.-J., and Lambelin, G. (1971). *J. Chromatogr.* **62,** 135.
Rosseel, M. T., and Bogaert, M. G. (1978). *J. Chromatogr.* **154,** 99.
Selley, M. L., Thomas, J., and Triggs, E. J. (1974). *J. Chromatogr.* **94,** 143.
Slack, J. A., and Ford-Hutchison, A. W. (1980). *Drug Met. Disposit.* **8,** 84.
Smith, D. A., and Cole, W. J. (1975). *J. Chromatogr.* **105,** 377.
Sparacino, C. M., Pellizzari, E. D., Cook, C. E., and Wall, M. W. (1973). *J. Chromatogr.* **77,** 413.
Strong, J. M., and Atkinson, A. J. (1972). *Anal. Chem.* **44,** 2287.
Testa, B., and Beckett, A. H. (1973). *J. Pharm. Pharmacol.* **25,** 119.
Thompson, R. M., and Gerber, N. (1976). *J. Chromatogr.* **124,** 321.
Trebbi, A., Gervasi, G. B., and Comi, V. (1975). *J. Chromatogr.* **110,** 309.

Vanderheeren, F. A. J., Theuwis, D. J. C. J., and Rosseel, M. T. (1976). *J. Chromatogr.* **120,** 123.
Verebely, K., and Inturrisi, C. E. (1973). *J. Chromatogr.* **75,** 195.
Wolen, R. L., and Gruber, C. M. (1968). *Anal. Chem.* **40,** 1243.
Yeh, S. Y., McQuinn, R. L., and Gorodetzky, C. W. (1977). *J. Pharm. Sci.* **66,** 201.
Zingales, I. A. (1971). *J. Chromatogr.* **61,** 237.

Chapter 8

Drug Screening

At first sight it might seem that gas chromatography is not a very promising method for drug screening since all that is measured is the time taken for a compound to travel from the injection port to the detector under a given set of conditions such as gas flow, column length, and stationary phase. However, the fact that many drugs have similar retention times can be used to advantage in screening, since it is often important to be able to say that a particular drug is not present in significant amounts.

A drug screen may be necessary in a variety of situations: an unconscious youth may be found in the street, someone may have died suddenly under suspicious circumstances, or an athlete may have demonstrated a sudden improvement in performance. Although the principles of drug screening are the same in each case, there are a number of important differences in scope and emphasis, depending on whether the situation is clinical, forensic, or sporting. These will be considered in the following sections.

I. Emergency Clinical Screening

If someone is admitted to a hospital in a comatose state, drug overdosing must always be considered a possibility. It is essential that a screen be rapidly carried out, since treatment should be instituted as soon as possible. It is of no use to have a very sophisticated and elegant screening procedure that may take several days, since by then the patient will usually have died or recovered of his own accord. Similarly, it is of little use to screen for a number of drugs as an emergency if there is no effective treatment. However as has been said, to the clinician it can be just as helpful to exclude certain drugs as to confirm the presence of others.

Clinical drug screens are generally carried out on urine specimens because of the drug-concentrating effect of the kidneys (provided that several hours have elapsed since ingestion of the dose). Plasma, serum, or blood can also be used for a quantitative or semiquantitative estimate of the amount taken, and stomach washings can be invaluable because it is often possible to retrieve undigested parts of tablets.

Gas chromatography is only one of a range of tools that should be used in drug screening. Simple color tests on urine can be used to detect salicylates, paracetamol, phenothiazines, and related drugs, and thin-layer chromatography can yield a great deal of information within a half-hour. However there are many drugs that cannot be readily detected by simple color tests, and a number of tests become unreliable when more than one drug is present.

Irrespective of the analytical technique used, it is usually necessary to purify the drug or drugs partially, and this is most frequently done by a series of extractions with organic solvents at different hydrogen-ion concentrations. A typical scheme is illustrated in Fig. 8.1. It can be seen that a urine sample can be rapidly treated to give a basic fraction, an acid fraction, and a weak acid and neutral fraction. More elaborate extractions can be developed, but these are more time consuming, and speed is essential in an emergency clinical screen. More information on this area can be obtained from books by Clarke (1969), Curry (1969), and Sunshine (1969).

An early example of the use of gas chromatography for clinical screening is given in a paper by Proelss and Lohmann (1971) who produced a system for sedatives and tranquilizers. In their scheme the following classes of drug could be detected: barbiturates, carbamates, glutarimides, succinimides, higher alcohols, hydantoins, phenothiazines, benzodiazepines, and dibenzazepines. Serum was chosen as the fluid for analysis, and a solvent extraction was carried out to separate the mixture into three fractions containing acidic, neutral, and basic drugs. In this way serum can be rapidly extracted to yield three separate fractions, which can then be dried over sodium sulfate, evaporated, and redissolved in absolute ethanol for direct injection onto the appropriate column. A summary of the columns used with retention data for relevant drugs is given in Tables 8.1–8.3.

It can be seen that by chromatographing serum extracts under the conditions described a rapid screen for a wide range of drugs can be carried out. Shortly after the work by Proelss and Lohmann (1971), Watson and Kalman (1972) described a method capable of screening for sedatives in only 30 min. Using barbital as the internal standard, plasma was made acidic and extracted with chloroform. The organic phase was separated by centrifugation and evaporated to dryness; the residue was redissolved in a small volume of chloroform, and a portion was injected onto a glass column (6 ft × 2 mm)

A.

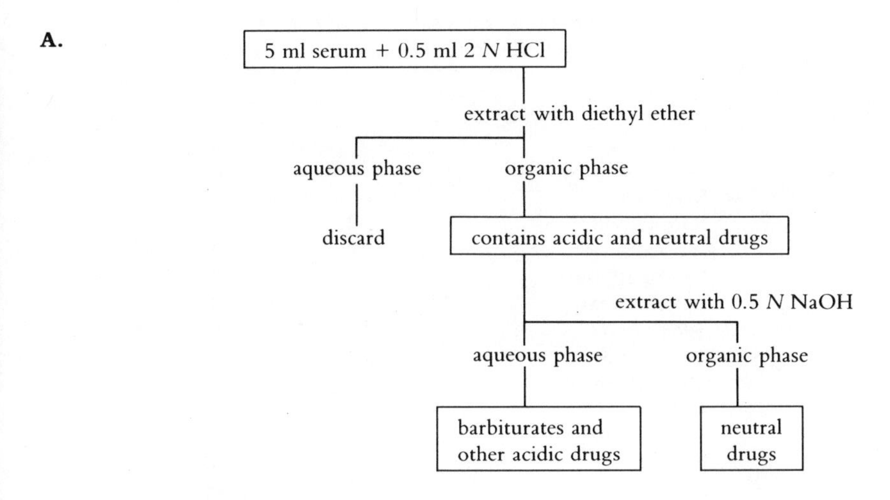

B.

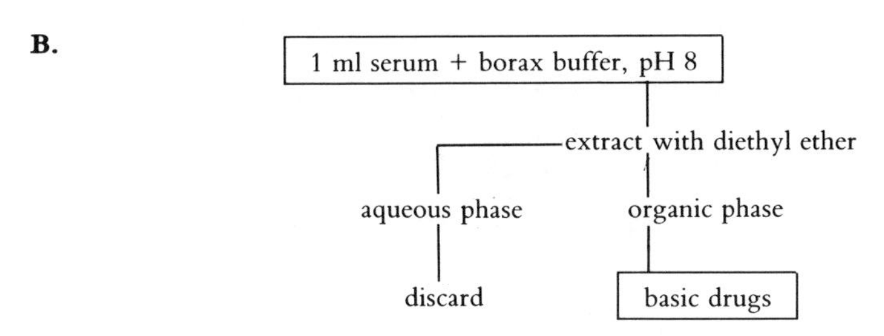

FIG. 8.1. Proelss and Lohmann extraction schemes for **(A)** acidic and neutral drugs and **(B)** basic drugs.

packed with 3% OV-17 and operated at 206°C. The drugs detected under these conditions are given in Table 8.4. In cases where the identification of a drug was doubtful, a portion of the chloroform extract was chromatographed on a second column of different polarity (3% PPE-20).

In 1972 Toseland *et al.* produced a simple method for monitoring therapeutic concentrations of a range of antiepileptic drugs: carbamazepine, diphenylhydantoin, phenobarbital, and primidone. Carbamazepine was extracted under alkaline conditions, while the other drugs were extracted from acidified plasma. To monitor carbamazepine, plasma was made alkaline and extracted with chloroform. The organic phase was separated by centrifuga-

Table 8.1

Retention Data for Barbiturates and Related Acidic Drugs[a]

Drug	Retention time	
	Absolute (min)	Relative
Barbital[b]	4.1	1.00
Butobarbital[b]	6.6	1.61
Amobarbital[b]	7.6	1.84
Pentobarbital[b]	8.4	2.03
Hexobarbital[b]	9.5	2.30
Secobarbital[b]	9.6	2.33
Thiopental[b]	9.0	2.18
Mephobarbital[b]	12.3	2.96
Allobarbital[b]	18.4	4.45
Heptabarbital[b]	22.4	5.42
Phenobarbital[b]	22.5	5.45
Ethchlorvynol[b]	20.5	5.00
Phenobarbital[c]	4.9	1.00
Primidone[c]	19.0	3.90
Diphenylhydantoin[c]	20.0	4.10

[a] After Proelss and Lohmann (1971). Column packing was 3.5% XE-60.

[b] Temperature programing: 12 min at 200°C, then a 20°C linear rise over 8 min. Helium flow rate 80 ml/min.

[c] Isothermal at 240°C, relative to phenobarbital.

Table 8.2

Retention Data for Sedatives and Related Neutral Drugs[a]

Drug	Retention time	
	Absolute (min)	Relative
Chloral hydrate	1.2	0.12
Oxanamide	1.4	0.13
Ethinamate	2.4	0.23
Styramate	2.5	0.24
Meprobamate	5.4	0.52
Methylprylon	6.4	0.62
Methsuximide	6.7	0.64
Mephenesin	6.9	0.66
Mebutamate	8.1	0.78
Tybamate	8.3	0.80
Methocarbamol	10.0	0.96
Phensuximide	10.4	1.00
Chlorphenesin carbamate	15.2	1.27
Glutethimide	19.3	1.86

[a] A column packed with 3.5% XE-60 was used and operated isothermally at 180°C, carrier gas flow 60 ml/min. From Proelss and Lohmann (1971).

Table 8.3

Retention Data for Tranquilizers and Antidepressants on Three Different Liquid Phases[a]

Drug	5% QF-1		3.5% XE-60		3% OV-17	
	A	B	A	B	A	B
Phenothiazine	6.5	1.00	4.25	1.00	4.9	1.00
Triflupromazine	8.0	1.23	2.50	0.59	5.4	1.10
Amitriptyline	—	—	—	—	5.9	1.20
Imipramine	6.0	0.92	2.25	0.53	6.3	1.28
Nortriptyline	—	—	—	—	6.8	1.39
Desipramine	6.8	1.04	3.00	0.71	7.6	1.56
Promazine	10.0	1.54	3.75	0.88	8.3	1.69
Oxazepam	15.8	2.42	—	—	12.0	2.46
Chlorpromazine	16.0	2.46	6.0	1.41	15.6	3.21
Diazepam	25.0	3.85	11.75	2.76	17.3	3.54
Mepazine	—	—	—	—	20.5	4.18
Trifluoperazine	22.0	3.38	9.0	2.12	22.5	4.62
Chlordiazepoxide	16.3	2.50	11.50	2.70	24.9	5.10
Thioridazine	—	—	35.00	8.24	23.0	4.71
Prochlorperazine	—	—	21.50	5.06	30.1	6.17

[a] A = absolute retention time (min), B = relative retention time. Operating conditions were as follows. QF-1 column: 210°C for 18 min, then programed ballistically to 240°C, nitrogen flow rate 100 ml/min. XE-60 column: isothermally at 235°C, helium flow rate 80 ml/min. OV-17 column: isothermally at 235°C, helium flow rate 60 ml/min. From Proelss and Lohmann (1971).

Table 8.4

Relative Retention Times of Sedatives[a]

Drug	Relative retention time
Barbital[b]	1.00
Butabarbital	1.87
Amobarbital	2.10
Pentobarbital	2.46
Secobarbital	2.93
Meprobamate	6.26
Glutethimide	6.77
Phenobarbital	8.50

[a] Detected by the method of Watson and Kalman (1972).

[b] Retention time, 1.08 min.

tion, evaporated to dryness, and the residue was redissolved in 0.05 *M* sulfuric acid. This was saturated with ammonium sulfate and extracted with diethyl ether. A chloroform solution of internal standard (perylene) was added to the ether, and the solution was evaporated to dryness. The residue was redissolved in 50 μl of methanol, then 2–5 μl was injected onto a column packed with 2% SP-1000 on Gas Chrom W HP operated at 250°C.

For the other drugs, plasma was acidified with 2 *M* formic acid and extracted twice with chloroform. The combined extracts were shaken with 1 *M* NaOH, and the separated aqueous phase was saturated with ammonium sulfate. This phase was then made acid with 0.5 ml concentrated HCl and extracted with diethyl ether. To the ether phase was added an internal standard (dibenzyl phthalate) in diethyl ether, and the solution was evaporated to dryness. The residue was redissolved in 50 μl of methanol, and 2–5 μl was injected onto the column described above. Because phenobarbital is often present in high concentration and gives a much higher FID response than the rest, the authors often found it necessary to prepare two dilutions so that adequate peaks could be obtained for primidone and diphenylhydantoin.

Varma (1978) published a method to monitor the same drugs as well as methsuximide and mephenytoin in the presence of Kemadrin and Prolixin, two drugs sometimes co-prescribed with these anticonvulsants. A single extraction scheme was outlined in which plasma was buffered at pH 6.5 with phosphate and then extracted with methylene chloride. The separated methylene chloride was evaporated to dryness, and the residue was redissolved in trimethylphenylammonium hydroxide, and a portion was injected onto a glass column packed with 3% OV-17 on Chromosorb W HP (60–80 mesh) operated at 140°C for 2 min and then programed to 230°C at a rate of 4°C/min. The drugs are converted on column to their methyl derivatives. The authors found that recoveries were greater than 98%, and the retention times, under the conditions described above, are given in Table 8.5. The internal standard 5-(*p*-methylphenyl)-5-phenylhydantoin had a retention time of about 17.5 min, while Kemadrin and Prolixin were eluted at 12.9 and 29.8 min, respectively.

These last two methods have been included to show that methods used for therapeutic drug monitoring can be used directly to screen for drugs in overdose. In many hospitals the therapeutic drug monitoring and overdose screening will often be carried out in the same department, sometimes by the same staff, and a wider screen can be carried out with very little extra effort. Mitchell *et al.* (1979) have used N-FID to screen for antidepressants in very small volumes of plasma (100 μl). No derivatization is needed, and recoveries are reported as being greater than 90%. Clomipramine was used as internal standard (2 μmol/liter), and 100 μl of this solution was evapo-

Table 8.5

Retention Times of Derivatized Anticonvulsants[a]

Drug	Retention time (min)
Methsuximide	3.8
Mephenytoin	6.9
Phenobarbital	7.8
Carbamazepine	10.5
Primidone	12.2
Diphenylhydantoin	16.2

[a] From Varma (1978).

rated to dryness in a test tube. Then 100 μl plasma was added, followed by an equal volume of 2 *M* NaOH. The tube was placed in a boiling-water bath for 5 min and then 600 μl of hexane was added with vortexing. The tube was placed in a water bath at 60°C for 2 min and then centrifuged for 15 min. The organic phase was removed and evaporated to dryness, and the residue was redissolved in 20 μl of methanol. A portion of this was injected onto a column of 3% OV-17 on Gas Chrom Q (100–120 mesh) operated at 225°C. The relative retention times are given in Table 8.6. Imipramine and nortriptyline were not separated on this column, but these are not normally prescribed to the same individual. Of course, in a drug overdose they may well be consumed together. The within-batch coefficient of variation was 9% or less.

Anyone engaged in emergency screening should make a point of reviewing the literature regularly and noting particularly any methods that describe the determination of a group of drugs rather than just individual members. For example, there is available a sensitive method, using derivati-

Table 8.6

Relative Retention Times of Antidepressants[a]

Drug	Relative retention time
Chlordiazepoxide	0.42
Amitriptyline	0.50
Trimipramine	0.50
Imipramine	0.55
Nortriptyline	0.59
Desipramine	0.66
Clomipramine	1.00
Chlorpromazine	1.44

[a] From Mitchell *et al.* (1979).

zation and electron capture that will allow the screening of five common antidiabetic drugs: chlorpropamide, tolbutamide, carbutamide, tolazamide, and glycodiazin (Schlicht *et al.,* 1978).

II. Forensic Screening

In forensic drug screening, although the principles are similar to emergency clinical screening, there are several important differences. In emergency screening, for example, the number of drugs taken in overdose for which specific treatment is available is small. If these can be ruled out, the patient is generally given a stomach washout, and electrolytes and blood gases are monitored until he revives. In forensic cases where a death or other serious crime is involved, it is important that the drug screen be as wide ranging and as thorough as possible; speed of analysis takes second place.

Urine is often the first body fluid examined because of the concentrating power of the kidney, and if a death is involved, a range of tissue specimens will generally be available. As in clinical screening, a series of extractions with organic solvents can be used, and acidic, neutral, and basic fractions are produced. The preliminary treatment of forensic specimens is often more complex because of the nature of the tissues (brain, liver, etc.) and because putrefaction may have begun in the specimens.

A. Putrefaction of Tissues

Because relatively nonselective chromatographic columns are frequently used in forensic investigations, it is not surprising that a number of endogenous biological compounds will interfere. Early studies of these were made by Anders and Mannering (1967) and Niyogi and Rieders (1971). These authors showed that apparent barbiturate, tripelennamine, hydroxyglutethimide, and desipramine GC peaks could be mimicked by stearic acid, *p*-hydroxybenzaldehyde, and tryptamine. Stevens and Evans (1973) have reviewed this subject in detail and have gathered data on 29 organic bases that can be produced as the result of tissue putrefaction. As one would expect, GC information is not used in isolation, and a number of other techniques are used to produce a more complete picture. However, we shall concentrate on the gas chromatographic details. Stevens and Evans list alphabetically the bases likely to be encountered in a general extraction such as the Stas–Otto. However, they point out that the number of potentially interfering bases can be reduced if protein precipitation and solvent extraction are first carried out. Also included in this valuable paper is a list of compounds

Table 8.7

Potentially Interfering Bases Chromatographing on 2.5% SE-30[a]

Base	Column temp (°C)	Retention time (sec)
Isoamylamine	100	38
Isobutylamine	100	29
Cadaverine	100	125
Ethanolamine	100	37
Ethylamine	100	24
Methylamine	100	24
1-Phenylethylamine	100	160
2-Phenylethylamine	100	210
Piperidine	100	47
n-Propylamine	100	26
Putrescine	100	73
Pyridine	100	43
Pyrrolidine	100	36
Harman	200	255
Harmine	200	630
Nicotinamide	200	52
N-Methylnicotinamide	200	59
Norharman	200	255
Piperidine	200	46
Thymine	200	156
Tryptamine	200	135
Tyramine	200	58

[a] After Stevens and Evans (1973).

which, although not products of putrefaction, are present in human tissues: nicotinamide, acetylcholine, choline, neurine, histamine, and their salts. For gas chromatography a 5-ft × 1/8-in glass column was used and packed with 2.5% SE-30 on Chromosorb G AW-DMCS (80–100 mesh). Volatile bases were run at a column temperature of 100°C, while the less volatile were run at 200°C. Very polar bases such as choline and acetylcholine did not chromatograph. The retention times have been summarized in Table 8.7.

In 1980 Ramsey *et al.* produced retention index data, on an SE-30 or OV-1 column, of almost 300 nondrug substances likely to be encountered in toxicological analyses. As well as the compounds resulting from putrefaction, the list includes numerous plasticizers, antioxidants, food additives, flavors, and scintillation reagents. This is a most valuable collection of data and should be available for reference in all laboratories engaged in drug screening.

One approach to obtaining cleaner extracts for gas chromatographic analysis is to use preparative-scale GC before injection onto the analytical column. McAuley and Kofoed (1970) proposed this approach and collected fractions in specially designed bottles cooled by an acetone–dry ice mixture. Recoveries of drug were greater than 90% for barbiturates, glutethimide, and meprobamate, over 80% for chlordiazepoxide, and greater than 70% for some phenothiazines.

Alha and Korte (1972) successfully used membrane filtration to obtain cleaner extracts. In their study, aqueous extracts from postmortem blood were filtered through 0.22- or 0.45-μ filters (Millipore) without serious loss. The authors illustrated their paper with some interesting and impressive chromatograms (Fig. 8.2). Kauert *et al.* (1982) have suggested the use of thin-channel ultrafiltration. In this technique diluted blood or tissue homogenate is pumped round a spiral while simultaneously being subjected to ultrafiltration by pressure. The authors claim that the filtrate is generally clear, although in some cases filtrate from liver homogenate may be slightly yellow presumably due to bilirubin. The ultrafiltrates can easily be further treated by liquid–liquid or liquid–solid extraction.

Tye and Freitag (1980) have published a simple method that produces essentially lipid-free drug extracts from blood. Blood is first extracted with chloroform, which is then evaporated to dryness. The residue is dissolved in

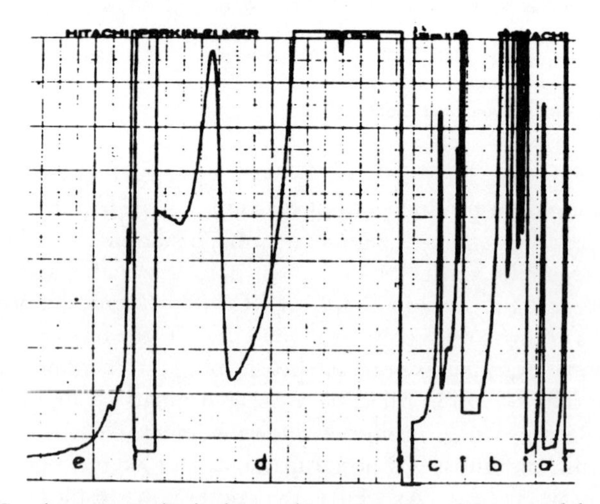

FIG. 8.2. Gas chromatographic background investigation. a, 0.5 μg amylobarbitone in 0.5 μl $CHCl_3$; b, corresponding amounts of amylobarbitone added to blood, isolation without Millipore filtration, many impurity peaks; c, as above with Millipore filtration, amylobarbitone peak detectable; d, a negative drug case, putrified blood, analysis without Millipore filtration; e, as above with Millipore filtration. From Alha and Korte (1972).

a small volume of petroleum ether and shaken with 8 volumes of 5% HCl. The authors claim that acidic, basic, and neutral drugs are successfully extracted into the aqueous phase, leaving fats behind in the petroleum ether. A reduction in lipid content of 1000-fold is claimed with drug recoveries generally greater than 50%.

An alternative approach is to remove the protein before solvent extraction, and this is usually essential when tissue homogenates such as brain, liver, and kidney are being examined. If this is not done, the formation of emulsions can be particularly troublesome. There are a number of methods of precipitating protein from biological fluids ranging from some of the earliest, using tungstic acid (Valov, 1946), ammonium sulfate and dilute acid (Nicholls, 1956), and acidic aluminum chloride (Stevens, 1967) to the enzymatic method (Osselton, 1977). Some of these are better suited to one group of drugs than another: for example, the Valov method is better for acidic and neutral compounds, while the enzymatic procedure gives better recoveries with basic drugs. Although the Nicholls method produces relatively clean extracts with basic drugs, recoveries are low. The enzymatic approach will now be examined in more detail.

In the enzymatic method, tissue samples are homogenized with 1 *M* Tris buffer (pH 10.5) and then incubated for 1 h at 50–60°C with a few milligrams of subtilisin Carlsberg. The solution is then cooled, filtered, and extracted with diethyl ether. This ether extract can then be used for a variety of chromatographic and spectroscopic analyses (Osselton, 1977; Osselton *et al.,* 1978). Drug concentrations that were measured by following this technique were compared with those following acid hydrolysis, and in general, much higher concentrations were obtained using the former method. Gas chromatography was the analytical technique used. However, it is worth noting that Dunnett and Ashton (1980), while finding the enzymatic approach useful for basic drugs, were unable to achieve satisfactory results with acidic or neutral compounds, and they continued using the tungstic acid method for these.

B. Identifying Unknowns

After a series of extractions have been carried out on a biological fluid or tissue homogenate, the resulting acidic, basic, and neutral extracts are usually injected onto GC columns of low to moderate polarity. Any peaks that appear can be partially identified by their retention indices. This approach has been outlined in Chapter 1, and it has the advantage that a more accurate identification can be obtained than is possible by using retention times relative to some standard substance.

An early illustration of this approach is given by Kazyak and Permisohn (1970) who produced tables of retention indices for many commonly used drugs and pesticides on three different stationary phases: OV-1, OV-17, and QF-1. Others have adopted this approach, most notably Moffat and colleagues, who published a very important series of papers. In one of the earlier ones (Moffat *et al.*, 1974) eight stationary phases of widely different polarity were compared in an effort to identify phases most useful for basic drug identification. The phases studied were SE-30, Apiezon L/KOH, OV-17, Carbowax 20M/KOH, Carbowax 20M, DEGS/KOH, and DEGS. The phases of lowest polarity were found to be the most discriminating, namely, SE-30 and OV-17. Many of the drugs could not be eluted from some of the more polar columns. For basic drug identification the authors recommended the use of either SE-30 or OV-17. This work culminated in a study of the GC retention indices of over 1300 drugs and other compounds of toxicological interest on SE-30 and OV-1 (Ardrey and Moffat, 1981). This most valuable paper should be in the possession of anyone seriously concerned with the use of GC as a screening tool for drugs.

Of course, the retention index alone does not identify a drug, and other analytical techniques such as UV and IR spectroscopy, TLC, and color tests may also have to be used. However the retention index approach is the most efficient means of narrowing down a range of possibilities and of confidently excluding others. Dutt (1982) has attempted a rather different approach and tried to identify common drugs by the multiple peaks some produce on gas chromatography or by the multiple peaks of their trimethylsilyl derivatives. Of the drugs he examined, 46 out of 116 gave more than a single peak. A 3% OV-17 column was used for this study, and Dutt claimed that it was possible to identify drugs by using their multiple peak characteristics.

Once the drug has been identified, it is then quantitated and the next question to be answered is, "What does the result mean?". A detailed discussion of this is outside the scope of this book, but the reader should bear in mind that many drugs exhibit a wide intersubject variability even in controlled clinical studies in young, healthy volunteers. Hence it is very difficult to predict with confidence from a single blood or urine concentration how much of a drug has been ingested. When, as is often the case, several drugs have been taken at the same time, the problem becomes even more difficult. However, bearing these qualifications in mind, it is possible to produce a guide to therapeutic, toxic, and fatal concentrations. The interested reader is referred to the work by Winek (1976), Baselt (1978), Pribor *et al.* (1980), and the compilation by Stead and Moffat (1983).

III. Drugs in Sport

There exists an ever increasing need to screen rapidly urine specimens obtained from athletes at major sporting events such as the Olympic Games and World Cup. Speed of analysis is essential but so too is certainty of identification. In an early paper in this field, Ariëns (1964) described long- and short-term conditioning with drugs. Long-term conditioning involves the use of hormones and anabolic steroids over a sustained period, and it is now generally recognized that this type of conditioning can very seriously damage the health of the athlete. Short-term conditioning involves the ingestion of stimulant drugs, and this is a common practice among some athletes. However, it should not be forgotten that in other sports, such as dog and horse racing, sedative drugs may be administered to a favorite in order to improve the chances of a less fancied runner.

Amphetamines and related compounds are the commonest examples of drugs used to improve performance, but new drugs are being produced each year, and the analyst should do his best to be aware of these. One of the earliest papers in this field is by Beckett and colleagues (1967) in which factors are reviewed that are of importance regarding the excretion of ionized drugs. For example, exercise produces an acidosis, which can be offset by the ingestion of sodium bicarbonate. As well as counteracting to some extent the fatigue, this will also render the urine of the athlete alkaline and reduce the excretion of any basic compounds. If he has taken a stimulant, such as amphetamine, its excretion will be reduced, and consequently its effect prolonged. Beckett and colleagues used four types of column packing, three incorporating alkali because of the basic nature of the drugs being sought. The stationary phases used were 5% Carbowax 6000 plus 5% KOH, 2% Carbowax 20M plus 5% KOH, 10% Apiezon L plus 10% KOH, and 2.5% SE-30. The first two columns were most often used in detecting the low molecular weight drugs such as the amphetamines. The Carbowax 20M, Apiezon L, and SE-30 columns were used for the higher molecular weight drugs.

The method was able to detect 0.1 μg drug per ml urine, but GC of the unchanged drug was not used in isolation, however, and derivative formation and TLC were also employed. Schiff bases with ketones, silyl derivatives, or fluorinated acyl derivatives were employed as appropriate. The sensitivity of the method was sufficient to allow detection of a stimulant drug in urine up to 48 h after ingestion of a normal dose.

In a paper published in 1974, Donike and Stratman describe the screening procedure used at the 1972 Olympic Games in Munich. The authors used a 1.06-m × 2.5-mm glass column packed with alkali-treated AW Chromosorb W (60–80 mesh) that had been coated with 2% Igepal CO-880 and 12.5% Apiezon L. The column was operated in a temperature-programed mode from 130 to 270°C at 2°C/min with N-FID. Diphenylamine was used as the internal standard, and 10 nitrogen-containing drugs were separated: heptaminole, amphetamine, methamphetamine, dimethylamphetamine, nicotine, ephedrine, phenmetrazine, nikethamide, pentamethylenetetrazole, and caffeine.

IV. Quality Control in Drug Screening

The clinical chemists have done much to pioneer quality control in laboratories, and several excellent books are available that outline the principles in detail (Whitehead 1977; Ottaviano and Di Salvo 1977). Quality control procedures should be carried out within each laboratory, and comparisons between laboratories can also be extremely useful. Although the subject is too complex to treat here, the general experience where quality control schemes have been instituted is that a few laboratories do well, some are satisfactory, and a number perform extremely badly. Continued participation in the scheme usually leads to a general improvement in the poorest laboratories, and if organized correctly, a healthy competitive instinct can develop as each laboratory seeks to improve its overall position. Toxicologists came rather late to quality control, but as early as 1973 Sunshine had published an exhortation entitled "Stand Up and Be Counted." In this he stressed the tremendous responsibility placed on the shoulders of each toxicologist, since his results could have a very serious effect on the individual whether he was an athlete, poison victim, parolee, or even a job applicant. He cited the results of a survey conducted by the Center for Disease Control where 163 laboratories from all over the United States took part in a screening program. Less than half of the laboratories (72) analyzed all specimens correctly, there were 110 false positives (which included 22 for morphine and 42 for methadone), and 153 false negatives (which included 33 for morphine and 10 for methadone).

Dinovo and Gottschalk (1976a) published a report of a survey of toxicology proficiency in nine laboratories. The testing program was set up to help the National Institute of Drug Abuse and consequently was concerned with drugs such as barbiturates, morphine, methadone, methaqualone, benzodiazepines, and alcohol. Gas chromatography was the most frequently used analytical tool, and of course, differences could be anticipated where different analytical methods were employed. Five standard drug samples were sent to each laboratory, and each contained secobarbital (6.5 mg/liter) to measure the variations in this assay with time. Three of the samples were drug-free urine to which drugs had been added and two were aqueous solutions of human albumin similarly treated. Drug concentrations were chosen that would be "associated with low toxicity to provide a moderate challenge to the toxicological methodologies." Samples were lyophilized before dispatch, and the amount of information supplied with each was deliberately varied to simulate real working conditions. For example, in some cases it was stated that the sample contained barbiturates, while in others the actual drug was specified. In some cases no information at all was given. The study found that there were wide variations in precision and accuracy between laboratories, but these seemed to improve when increased information was provided. False negatives were again a problem. The authors concluded that the majority of laboratories responded well if asked to assay a specific drug in a sample but responded poorly if a screen had to be performed.

The study was criticized by Kelly and Sunshine (1976) on a number of grounds. For example, the composition of some of the spiked samples was claimed to be extremely unlikely, and some of the drugs were present in the unchanged form when usually they would be in the form of metabolites, including conjugates. Dinovo and Gottschalk, however, were able to defend themselves against many of the criticisms (1976b).

A detailed laboratory and method comparison has appeared from the New York State Department of Health (Buhl *et al.*, 1978). Barbiturates, phenytoin, procainamide, and glutethimide were the drugs used, and gas chromatography, UV spectrophotometry, fluorimetry, colorimetry, and immunoassay the methods compared. The report discussed a number of difficulties and sources of error. The authors concluded that proficiency-testing programs could lead to increased accuracy, but improvements should begin with internal quality control.

In a 1981 study, Ingelfinger *et al.* compared toxicology screening in three commercial laboratories with one academic laboratory. All the laboratories were well equipped, and each possessed a GC–MS among its instrumentation. All four laboratories agreed on only 6 of the 23 drugs reported as being present, and large differences in concentration were observed. For example, all reported secobarbital as being present in one case, the commercial laboratories finding concentrations of 4, 11, and 23.5 mg/liter, while the research laboratory reported a concentration of 5 mg/liter. In another case all reported phenytoin as being present, the commercial laboratories finding concentrations of 2.5, 20.5, and 26.0 mg/liter, while the research laboratory reported only a trace.

All of the studies reviewed above are American, and there does not appear to be a similar study published elsewhere, at least in the area of toxicological screening. It is unlikely that results from other countries would be appreciably better. Clearly there is much room for improvement.

References

Alha, A., and Korte, T. (1972). *Ann. Med. Exp. Biol. Fenniae* **50,** 175.

Anders, M. W., and Mannering, G. J. (1967). *Prog. Chem. Tox.* **3,** 121.

Ardrey, R. E., and Moffat, A. C. (1981). *J. Chromatogr.* **220,** 195.

Ariëns, E. J. (1964). *In* "Doping—Proceedings of an International Seminar" (A. De Schaepdryver and M. Hebbelinck, eds.), p. 27. Pergamon, London.

Baselt, R. C. (1978). "Disposition of Toxic Drugs and Chemicals," vols. 1 and 2. Biomedical Publications, Connecticut.

Beckett, A. H., Tucker, G. T., and Moffat, A. C. (1967). *J. Pharm. Pharmacol.* **19,** 273.

Buhl, S. N., Kowalski, P., and Vanderlinde, R. E. (1978). *Clin. Chem.* **24,** 442.

Clarke, E. G. C. (1969). "Isolation and Identification of Drugs." Pharmaceutica Press, London.

Curry, A. S. (1969). "Poison Detection in Human Organs," 2nd Edition. Thomas, Springfield, Illinois.

Dinovo, E. C., and Gottschalk, L. A. (1976a). *Clin. Chem.* **22,** 843.

Dinovo, E. C., and Gottschalk, L. A. (1976b). *Clin. Chem.* **22,** 2056.

Donike, M., and Stratman, M. (1974). *Chromatographia* **7,** 182.

Dunnett, N., and Ashton, P. G. (1980). *In* "Forensic Toxicology, Proceedings of the European Meeting of the International Association of Forensic Toxicology" (J. S. Oliver, ed.), p. 272. Croom Helm, London.

Dutt, M. C. (1982). *J. Chromatogr.* **248,** 115.

Ingelfinger, J. A., Isakson, G., Shine, D., Costello, C. E., and Goldman, P. (1981). *Clin. Pharmacol. Ther.* **29,** 570.

Kauert, G., Drasch, G., von Meyer, L., and Schneller, F. (1982). *In* "Proceedings of the European Meeting of the International Association of Forensic Toxicologists" (J. S. Oliver, ed.), p. 71. Croom Helm, London.

Kazyak, L., and Permisohn, R. (1970). *J. Forensic Sci.* **15,** 346.

Kelly, R. C., and Sunshine, I. (1976). *Clin. Chem.* **22,** 1413.

McAuley, F., and Kofoed, J. (1970). *J. Chromatogr.* **50,** 513.

Mitchell, W. D., Webb, S. F., and Padmore, G. R. A. (1979). *Ann. Clin. Biochem.* **16,** 47.

Moffat, A. C., Stead, A. H., and Smalldon, K. W. (1974). *J. Chromatogr.* **90,** 19.

Nicholls, L. C. (1956). "The Scientific Investigation of Crime." Butterworths, London.

Niyogi, S. K., and Rieders, F. (1971). *Acta Pharmacol. Toxicol.* **29,** 113.

Osselton, M. D. (1977). *J. Forensic. Sci. Soc.* **17,** 189.

Osselton, M. D., Shaw, I. C., and Stevens, H. M. (1978). *Analyst* **103,** 1160.

Ottaviano, P. J., and Di Salvo, A. F. (1977). "Quality Control in the Clinical Laboratory." University Park Press, Baltimore, Maryland.

Pribor, H. C., Morrel, G., and Scherr, G. H. (eds.). (1980). "Drug Monitoring and Pharmacokinetic Data." Pathotox Publishers, Illinois.

Proelss, H. F., and Lohmann, H. (1971). *Clin. Chem.* **17,** 222.

Ramsey, J. D., Lee, T. D., Osselton, M. D., and Moffat, A. C. (1980). *J. Chromatogr.* **184,** 185.

Schlicht, H.-J., Gelbke, H.-P., and Schmidt, G. (1978). *J. Chromatogr.* **155,** 178.

Stead, A. H., and Moffat, A. C. (1983). *Human Toxicology* **3,** 437.

Stevens, H. M. (1967). *J. Forensic Sci. Soc.* **7,** 184.

Stevens, H. M., and Evans, P. D. (1973). *Acta Pharmacol. Toxicol.* **32,** 525.

Sunshine, I. (1969). "Handbook of Analytical Toxicology." Chemical Rubber Co., Cleveland, Ohio.

Sunshine, I. (1973). *J. Chromatogr.* **82,** 125.

Toseland, P. A., Grove, J., and Berry, D. (1972). *Clin. Chim. Acta* **38,** 321.
Tye, R., and Freitag, J. (1980). *J. Forensic Sci.* **25,** 95.
Valov, P. (1946). *Ind. Eng. Chem. Anal. Ed.* **18,** 456.
Varma, R. (1978). *J. Chromatogr.* **155,** 182.
Watson, E., and Kalman, S. M. (1972). *Clin. Chim. Acta* **38,** 33.
Whitehead, T. P. (1977). "Quality Control in Clinical Chemistry." Wiley, New York and London.
Winek, C. L. (1976). *Clin. Chem.* **22,** 832.

Envoi: Enlightenment *by G. Machata*

There appears a colleague, wearing a trustful expression who says: "Since you are so well equipped with gas chromatographs, could you not . . .". The following scene is almost always the same and happens like this everywhere in the world. Some turbid fluid is presented, certainly the result of long and painstaking work, but wholly unsuitable for analysis. Now you have to enrich, to purify, to fractionate, to form derivatives, to study the literature, to carry out preliminary chemical tests, and many other things. It goes without saying that the highly recommended column packing material X is not available; will the available material Y also be suitable? Where are reference substances for identification? Why is it that, according to the literature, the temperature can be programmed up to 300°C without any baseline drift, while you achieve this no farther than up to 210°C? But finally everything works out, or almost everything. The gas chromatogram has been produced, the internal standard has been correctly chosen with regard to retention time and quantity, the temperature program rate provides optimum resolution, almost every peak can be identified by means of reference substances or literature information. You are satisfied, the colleague beams and exits overjoyed. But only a short time later, there is another knock or a ring at the telephone: "My dear colleague, could you not . . .".

Of course you could and you should; but why is it that the fundamental requirements for a gas chromatographic analysis have to be newly stipulated each time? Never has it happened to me so far that somebody has come to me and said: "Please make a GC analysis of this liquid which contains 10% of a fraction in solution. I am interested in the ratio of the C_6 through C_{10} aldehydes. Here are the tests with pure substances mixed in the ratio that can be expected". I am still waiting for such a visitor.

Reprinted with permission from *Chromatographia* **3,** 52 (1970).

Similarly this is also true for all the other physical-chemical separation procedures. Is it possible that the reason for an increasing incomprehension is insufficient analytical instruction? Where is the non-analyst who does not imagine that there are automatic instruments where the sample is fed in and the complete analysis comes out at the other end? Who ever thinks of sample preparation work which can take much more time than the analysis itself? Are we to be degraded to act as "auxiliary scientists"? This attitude must be criticized particularly in the colleagues of neighbouring sciences, such as medical people who count increasingly among our "customers", although it is more excusable in them than in chemists.

Enlightenment is necessary. Let us start with instruction in analytical work. And let us start today.

Index

A

B

C

F

G

H

I

K

L

M

N

Q

R

S

T

U

V

W